Fractional Order PID and ADR Controls

Pengchong Chen and Ying Luo

CRC Press
Taylor & Francis Group
Boca Raton London New York

CRC Press is an imprint of the
Taylor & Francis Group, an **informa** business

Designed cover image: Getty Images

First edition published 2025
by CRC Press
2385 NW Executive Center Drive, Suite 320, Boca Raton FL 33431

and by CRC Press
4 Park Square, Milton Park, Abingdon, Oxon, OX14 4RN

CRC Press is an imprint of Taylor & Francis Group, LLC

ISBN: 978-1-032-50240-3 (hbk)
ISBN: 978-1-032-50261-8 (pbk)
ISBN: 978-1-003-39763-2 (ebk)

DOI: 10.1201/9781003397632

Typeset in Nimbus Roman
by KnowledgeWorks Global Ltd.

Contents

Part III Fractional Order ADR Control

Part IV Fractional Order PID and ADR Control

Preface

The ever-evolving field of control systems continues to push the boundaries of technology and innovation. As practitioners and researchers seek more robust and versatile solutions, Fractional Order Proportional-Integral-Derivative (FOPID) and active disturbance rejection (ADR) controls have emerged as significant advancements. This monograph, *Fractional Order PID and ADR Controls*, is born out of a desire to explore and elucidate these cutting-edge techniques.

The motivation of this monograph stems from the increasing complexity of modern control systems and the need for advanced methods to address these challenges. Fractional calculus offers a powerful extension to classical control theories, while ADR control provides a proactive approach to disturbance management. Together, these methods hold the promise of enhanced performance and reliability in a variety of control applications.

The primary goal of this monograph is to provide a comprehensive resource that drives the performance enhancement and bridges the gap between theory and practice in FOPID and ADR controls. By detailing the mathematical foundations, design principles, and practical implementations, we aim to equip readers with the knowledge and tools necessary to tackle complex control problems. Nonetheless, fractional order controllers encounter challenges in parameter tuning due to the superfluous fractional order parameters, while the application of active disturbance rejection control faces issues regarding its integration with actual systems.

This monograph is organized as follows. Part I contains Chapter 1, introducing fundamentals of fractional order PID and ADR controls followed by research motivations and contributions of this monograph. Part II which includes Chapters 2–5 focuses on the fractional order PID control. Part III, including Chapters 6–8, is dedicated to the fractional order ADR control. The fractional order PID and ADR control is presented in Part IV.

It is my sincere hope that this monograph will serve as a valuable resource in your study and application of FOPID and ADR controls. May it inspire you to explore new possibilities and drive innovations in the field of control systems.

Authors

Pengchong Chen earned a PhD in mechanical science and engineering at Huazhong University of Science and Technology, Wuhan, China, in 2022. He is currently an Associate Professor in the School of Electrical and Information Engineering at Zhengzhou University, Zhengzhou, China. Dr. Chen's research interests include mechatronics, robotic control, fractional order, and active disturbance rejection controls.

Ying Luo earned a PhD in control theory and engineering with a joint-PhD program at South China University of Technology (China) and Utah State University (USA) in 2009. He was an Assistant Research Professor at the Center for Self-Organizing and Intelligent Systems in the Department of Electrical and Engineering at Utah State University from 2009 to 2010. Dr. Luo is currently a Professor at the School of Automation Science and Engineering at the South China University of Technology, Guangzhou, China. His research interests include applied fractional calculus in modeling and controls, mechatronics and robotics, active disturbance rejection controls, and smart control engineering.

Acknowledgments

This book provides a comprehensive summary of our research efforts during the past few years in fraction order proportional integral derivative and active disturbance rejection controls theory. This book contains material from papers and articles that have been previously published as well as the Ph.D. dissertation of the first author. We are grateful and would like to acknowledge the copyright permissions from the following publishers who have released our works.

Acknowledgment is given to the Institute of Electrical and Electronics Engineers (IEEE) to reproduce material from the following papers:

2009 IEEE. Reprinted, with permission, from YangQuan Chen, I. Petras, and Dingyu Xue, Fractional order control: A tutorial, in *Proceedings of American Control Conference*, 10-12 June 2009, St. Louis, MO, pages 1397–1411 (material found in Chapter 1). doi: 10.1109/ACC.2009.5160719.

2002 IEEE. Reprinted, with permission, from Dingyu Xue, and YangQuan Chen, A comparative introduction of four fractional order controllers, in *Proceedings of the 4th World Congress on Intelligent Control and Automation*, 2002, pages 3228–3235 (material found in Chapter 1). doi: 10.1109/WCICA.2002.1020131.

2023 IEEE. Reprinted, with permission, from Pengchong Chen and Ying Luo, Analytical fractional-order PID controller design with Bodes ideal cutoff filter for PMSM speed servo system, in *IEEE Transactions on Industrial Electronics*, vol. 70, no. 2, pp. 1783–1793, Feb. 2023 (material found in Chapter 5). doi: 10.1109/TIE.2022.3158009.

2023 IEEE. Reprinted, with permission, from Pengchong Chen and Ying Luo, A two-degree-of-freedom controller design satisfying separation principle with fractional-order PD and generalized ESO, in *IEEE/ASME Transactions on Mechatronics*, vol. 27, no. 1, pp. 137–148, Feb. 2022 (material found in Chapter 10) doi: 10.1109/TMECH.2021.3059160.

2024 IEEE. Reprinted, with permission, from Pengchong Chen, Ying Luo, He Gan, Yanhong Liu and Yangquan Chen, A current- and speed-loop decoupling controller for SPMSM under periodic disturbances, in *IEEE Transactions on Power Electronics*, vol. 39, no. 6, pp. 6889–6902 (material found in Chapter 8) June 2024, doi: 10.1109/TPEL.2024.3377228.

Acknowledgment is given to the Elsevier B.V. to reproduce material from the following papers:

2021 Elsevier B.V. Reprinted, with permission, from Pengchong Chen, Ying Luo, Yibing Peng, and YangQuan Chen, Optimal robust fractional order PIλD controller synthesis for first order plus time delay systems, *ISA Transactions*, vol. 114, pp. 136–149 (material found in Chapter 3) Aug. 2022. doi: 10.1016/j.isatra.2020.12.043.

2022 Elsevier B.V. Reprinted, with permission, from Pengchong Chen and Ying Luo, An analytical synthesis of fractional order $PI^{\lambda}D^{\mu}$ controller design, *ISA Transactions*, vol. 131, pp. 124–136 (material found in Chapter 4) Aug. 2022. doi: 10.1016/j.isatra.2022.04.0473.

2021 Elsevier B.V. Reprinted, with permission, from Pengchong Chen and Ying Luo, Fractional order active disturbance rejection control with the idea of cascaded fractional order integrator equivalence, *ISA Transactions*, vol. 114, pp. 359–369 (material found in Chapter 6) Aug. 2022. doi: 10.1016/j.isatra.2020.12.030.

Acknowledgment is given to the Wiley to reproduce material from the following papers:

2022 Wiley. Reprinted, with permission, from Pengchong Chen, Weijia Zheng, Ying Luo, Yibing Peng and YangQuan Chen, Robust three-parameter fractional-order proportional integral derivative controller synthesis for permanent magnet synchronous motor speed servo system, *Asian Journal of Control*, vol. 24, no. 6, pp. 3418–3433 (material found in Chapter 2) May 2022. doi: 10.1002/asjc.2832.

Acknowledgment is given to the MDPI to reproduce material from the following papers:

2021 MDPI . Reprinted, with permission, from Pengchong Chen, Ying Luo, Yibing Peng, Yangquan Chen. Optimal fractional-order active disturbance rejection controller design for PMSM speed servo system. *Entropy* vol. 23, no. 3, pp. 262 (material found in Chapter 9), 2021. doi: 10.3390/e23030262.

List of Figures

List of Tables

Part I

Fundamentals of Fractional Order PID and ADR Controls

1 Introduction

The field of control systems has continually evolved, with new methodologies and technologies emerging to address complex and dynamic systems. Among these, fractional order proportional integral derivative (FOPID) and active disturbance rejection (ADR) controls have garnered significant attention for their robustness and versatility. The FOPID control extends the traditional PID controller by incorporating fractional calculus, allowing for more flexible and precise control strategies. This approach provides enhanced performance in terms of stability and response time, especially in systems with intricate dynamics that are not well-served by integer order models. The ADR control focuses on real-time estimation and compensation of disturbances within a system. By actively identifying and mitigating these disturbances, the ADR control ensures that the system maintains desired performance even in the presence of internal uncertainties and external disturbances.

1.1 FRACTIONAL CALCULUS

The fractional-order integral-differential operator is defined as follows [1]

$$
{}_{t_0}D_t^{\alpha} \begin{cases} \frac{d^{\alpha}}{dt^{\alpha}}, & \text{Re}(\alpha) > 0 \\ 1, & \text{Re}(\alpha) = 0 \\ \int_{t_0}^{t} (d\tau)^{-\alpha}, & \text{Re}(\alpha) < 0 \end{cases} \tag{1.1}
$$

where t_0 and t are the start time and the end time of the integration, respectively. The term α is the fractional order. $Re(\alpha)$ is the real part of α.

1.1.1 Definitions

To date, various definitions of fractional calculus have been proposed, and the commonly used definitions are summarized as follows [1]:

A. Grünwald–Letnikov definition

The fractional order differentiation and integral can be defined as

$$
{}_{t_0}^{GL}D_t^{\alpha} f(t) = \lim_{h \to 0} \frac{1}{h^{\alpha}} \sum_{j=0}^{[(t-t_0)/h]} (-1)^j \binom{\alpha}{j} f(t - jh), \tag{1.2}
$$

where GL notation signifies the Grünwald–Letnikov definition; the $\binom{\alpha}{j}$ are the binomial coefficients; the subscripts to the left and right of D are the lower and upper bounds in the integral. The value of α can be positive or negative, corresponding to differentiation and integration, respectively, and the α is non-integer.

DOI: 10.1201/9781003397632-2

B. Riemann–Liouville definition

The fractional order integral can be directly extended from the conventional integer order integral as

$$ {}_{t_0}^{RL}D_t^{-\alpha} f(t) = \frac{1}{\Gamma(\alpha)} \int_{t_0}^{t} (t-\tau)^{\alpha-1} f(\tau)\mathrm{d}\tau, \tag{1.3} $$

RL notation signifies the Riemann–Liouville definition; $0 < \alpha < 1$ is the fractional order. The fractional order differentiation is defined as

$$ {}_{t_0}^{RL}D_t^{\beta} f(t) = \frac{\mathrm{d}^n}{\mathrm{d}t^n}\left[{}_{t_0}^{RL}D_t^{-(n-\beta)} f(t)\right] = \frac{1}{\Gamma(n-\beta)} \frac{\mathrm{d}^n}{\mathrm{d}t^n}\left[\int_{t_0}^{t} \frac{f(\tau)}{(t-\tau)^{\beta-n+1}}\mathrm{d}\tau\right], \tag{1.4} $$

where $n-1 < \beta \leq n$.

C. Caputo definition

In cases where nonzero initial conditions are present, the description of the differential equation system may pose challenges. Therefore, in the study and simulation of nonzero systems, a certain solution to the conflict between the well-established and polished mathematical theory and practical needs was proposed by Caputo. The Caputo fractional order differentiation is defined by

$$ {}_{t_0}^{C}D_t^{\alpha} f(t) = \frac{1}{\Gamma(1-\gamma)} \int_{t_0}^{t} \frac{f^{(m+1)}(\tau)}{(t-\tau)^{\gamma}}\mathrm{d}\tau, \tag{1.5} $$

where $\alpha = m+\gamma$, m is an integer and $0 < \gamma \leq 1$. The Caputo fractional order integral is defined as

$$ {}_{t_0}^{C}D_t^{-\gamma} f(t) = \frac{1}{\Gamma(\gamma)} \int_{t_0}^{t} \frac{f(\tau)}{(t-\tau)^{1-\gamma}}\mathrm{d}\tau, \gamma > 0. \tag{1.6} $$

1.1.2 Stability of Fractional Order Systems

The fractional order linear time-invariant (LTI) system can be represented as following state-space model

$$ \begin{aligned} {}_0D_t^q x(t) &= Ax(t) + Bu(t), \\ y(t) &= Cx(t), \end{aligned} \tag{1.7} $$

where $x \in R^n$, $u \in R^r$ and $y \in R^p$ are the state, input, and output vectors of the system and $A \in R^{n\times n}$, $B \in R^{n\times r}$, $C \in R^{p\times n}$, q is the fractional commensurate order.

It is known from the theory of stability that a linear time-invariant system is stable if the roots of the characteristic polynomial are negative or have negative real parts if they are complex conjugate. It means that they are located on the left half of the complex plane. In the fractional order LTI case, the stability is different from the integer one. An interesting notion is that a stable fractional order system may have roots in the right half of complex plane. It has been shown that system (1.7) is stable if the following condition is satisfied [2]

$$|\arg(eig(A))| > q\frac{\pi}{2}, \tag{1.8}$$

where $0 < q < 1$ and eig(A) denotes the eigenvalues of matrix A.

Matignon's stability theorem says: The fractional transfer function $G(s) = Z(s)/P(s)$ is stable if and only if the following condition is satisfied in σ-plane:

$$|\triangle(\sigma)| > q\frac{\pi}{2}, \forall\sigma \in C, P(\sigma) = 0, \tag{1.9}$$

where $\sigma := s^q$. When $\sigma = 0$ is a single root of $P(s)$, the system cannot be stable. For $q = 1$, this is the classical theorem of pole location in the complex plane: no pole is in the closed tight half plane of the first Riemann sheet.

1.2 FOPID CONTROL

Bode [72] may have been one of the first to comprehensively consider the use of fractional order integral-differential operators in feedback loops, although the term "fractional" was not employed. Subsequently, Manabe [4] introduced the frequency-domain and transient responses of fractional-order integration and its application in control systems. In the further development of the field of automatic control, Lurie [5] proposed the tilt-integral-derivative (TID) controller in a patent, and Oustaloup introduced three generations of CRONE controllers [6–8], surpassing traditional integer order PID controllers and achieving superior performance. The generalization of the integral order PID controller was proposed by Podlubny [9], e.g., $\mathrm{PI}^{\lambda}\mathrm{D}^{\mu}$ controller with the fractional orders λ and μ.

The $\mathrm{PI}^{\lambda}\mathrm{D}^{\mu}$ was studied in the time domain in [10] and in the frequency domain in [9]. The transfer function of $\mathrm{PI}^{\lambda}\mathrm{D}^{\mu}$ is

$$C(s) = K_p + T_i s^{-\lambda} + T_d s^{\mu}, \tag{1.10}$$

where λ and μ are the fractional orders, K_p, T_i, and T_d are the proportional gain, integration constant and differentiation constant. If $\lambda = 1$ and $\mu = 1$, a classical PID is obtained. If μ, we obtain a PI^{λ} controller, etc. All these types of controllers are particular cases of the $\mathrm{PI}^{\lambda}\mathrm{D}^{\mu}$ controller.

Prior to this, various types of fractional order controllers had been proposed, including FOPI, FOPD, FO[PD], and so forth [11, 12]. It is worth noting that these controllers are all based on parameter tuning using three frequency-domain indices. However, there has been a lack of a systematic analytical design approach for a complete five-parameter FOPID controller. This book progressively increases the number of unknown parameters in fractional order controllers and ultimately presents an analytical frequency-domain tuning method for a complete five-parameter FOPID controller, filling this gap in the literature.

1.3 ADRC

ADRC was first proposed by Han to serve as an alternative control to the classical PID [13]. The extended state observer (ESO), which is the core part of ADRC, estimates the internal uncertainties (the unmodeled dynamics, uncertain dynamics) and the external disturbances as a lumped term called the total disturbance, then compensate the total disturbance in real time. However, the structure is complex and needs to tune a bunch of parameters, which makes it difficult to use in practice. To overcome the difficulty, Gao proposed a linear ADRC where linear ESO and linear state feedback are used [14].

Assume the controlled plant has the following model

$$y^{(n)}(t) = b_0 u(t) + f(y, \dot{y}, ..., y^{(n-1)}, u, \dot{u}, ..., u^{(n)}, w, t), \tag{1.11}$$

where f is a combination of the unknown dynamics and the external disturbance of the plant, and denoted as the total disturbance.

To estimate f, assume that f is differentiable and let $h = \dot{f}$, then an extended state-space form of (1.11) is

$$\begin{cases} \dot{z} = A_e z + B_e u + E_e h \\ \hat{y} = C_e z, \end{cases} \tag{1.12}$$

with the state $z = [z_1 \quad z_2 \quad ...z_n \quad z_{n+1}]^T$, where

$$z_1 = y, z_2 = \dot{y}, ..., z_n = y^{(n-1)}, z_{n+1} = f, \tag{1.13}$$

and state-space matrices

$$A_e = \begin{bmatrix} 0 & 1 & 0 & \cdots & 0 \\ 0 & 0 & 1 & \cdots & 0 \\ \vdots & \vdots & \vdots & \ddots & \vdots \\ 0 & 0 & 0 & \cdots & 1 \\ 0 & 0 & 0 & \cdots & 0 \end{bmatrix}_{(n+1)\times(n+1)}, \quad B_e = \begin{bmatrix} 0 & 0 & \cdots & b_0 & 0 \end{bmatrix}^T_{(n+1)\times 1},$$
$$E_e = \begin{bmatrix} 0 & 0 & \cdots & 0 & 1 \end{bmatrix}^T_{(n+1)\times 1}, \quad C_e = \begin{bmatrix} 1 & 0 & 0 & \cdots & 0 \end{bmatrix}_{1\times(n+1)}. \tag{1.14}$$

In the ADRC framework, the ESO is applied to estimate the total disturbance f. The linear ESO is designed as

$$\begin{cases} \dot{\hat{z}} = A_e \hat{z} + B_e u + L_o(y - \hat{y}), \\ \hat{y} = C_e \hat{z}, \end{cases} \tag{1.15}$$

where L_0 is the observer gain vector

$$L_0 = [\beta_1 \quad \beta_2 \quad ... \quad \beta_n \quad \beta_{n+1}]^T. \tag{1.16}$$

When $A_e - L_o C_e$ is asymptotically stable, $\hat{z}_1(t)$, ...,$\hat{z}_n(t)$ will approximate $y(t)$ and its derivatives and $\hat{z}_{n+1}(t)$ will approximate the generalized disturbance f. The estimated total disturbance can be compensated as in the following feedback law [14]:

$$u(t) = \frac{k_1(r(t) - \hat{z}_1(t)) + ... + k_p(r^{n-1}(t) - \hat{z}_n(t))}{b_0} - \frac{\hat{z}_{n+1}(t)}{b_0}, \tag{1.17}$$

where $r(t)$ is the reference signal.

1.4 WHY FOPID AND ADR CONTROLS?

FOPID has been empirically demonstrated to achieve superior control performance in dynamic response or disturbance attenuation, owing to the flexibility introduced by fractional order differentiator and integrator. It can adapt to different systems to meet users' more nuanced demands for various performance criteria. Active disturbance rejection control has garnered widespread attention among scholars due to its remarkable disturbance rejection capability and adaptability. Its core concept involves grouping external disturbances, model uncertainties, and unmodeled dynamics into a total disturbance, which is then actively observed and compensated for in real-time using an extended state observer. Usually, an excellent control system should have optimal dynamic tracking performance and disturbance rejection performance, and fractional order controllers and active disturbance rejection controllers perform well in these two aspects, respectively. Therefore, in this book, we will focus on how to organically combine FOPID with ADR controls.

1.5 PMSM SERVO AND EXPERIMENTAL PLATFORM

1.5.1 PMSM Speed Servo Model

The general structure of the permanent magnet synchronous motor (PMSM) servo system based on vector control is shown in Figure 1.1. Assume that hysteresis and eddy current loss and the saturation nonlinear factor of the magnetization curve are ignored. The voltage equations of PMSM in the rotating reference d-q axis are expressed as follows:

$$\begin{aligned} \dot{i}_d &= -\frac{R}{L}i_d + \frac{1}{L}u_d + n_p\omega L i_q, \\ \dot{i}_q &= -\frac{R}{L}i_q + \frac{1}{L}u_q - \frac{n_p\omega}{L}(Li_d + \psi_f), \end{aligned} \tag{1.18}$$

where R and L are phase resistance and inductance, respectively, u_d and u_q are the d-axis and q-axis stator voltages, i_d and i_q are the d-axis and q-axis stator currents, ω is the motor angular velocity and n_p is the pole pairs number, and ψ_f is magnetic flux linkage.

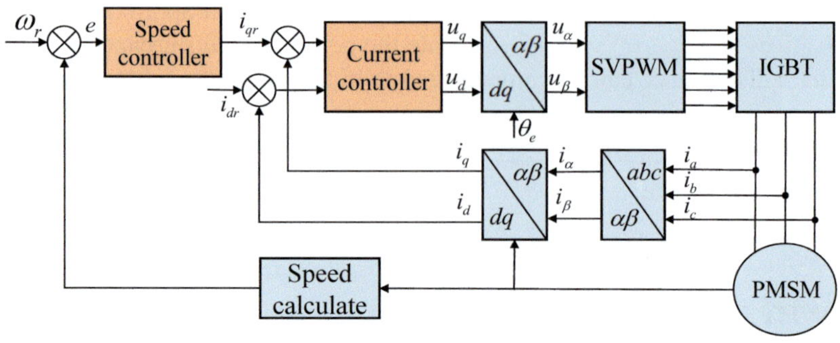

FIGURE 1.1 Block diagram of vector control of the PMSM.

With $i_d = 0$ control strategy in field-oriented control (FOC) framework, the dynamic model of q-axis can be presented as

$$\dot{i_q} = \frac{1}{L}u_q - \frac{R}{L}i_q. \tag{1.19}$$

The transfer function form can be expressed as

$$G_e(s) = \frac{1}{Ls+R}. \tag{1.20}$$

The mechanic equation of PMSM can be expressed

$$\begin{aligned} \dot{\omega} &= \frac{1}{J}T_e - \frac{1}{J}T_f - \frac{1}{J}T_L, \\ T_e &= 1.5n_p\psi_f i_q, \end{aligned} \tag{1.21}$$

where J is the moment of inertia constant. T_e, T_f, and T_L are the electromagnetic torque, friction torque, and external load torque, respectively.

The transfer function of the mechanical dynamic can be expressed as

$$G_m(s) = \frac{1.5n_p\psi_f}{Js}. \tag{1.22}$$

With the above analysis and Laplace transform, the block diagram of the PMSM dual-loop control system is shown in Figure 1.2.

1.5.2 Experimental Platform

The experimental platform is shown in Figure 1.3, which is composed of a PMSM, a DC generator, a servo drive, and a PC. The model of the motor is VM7-M13A-2R020-D1 and parameters are listed in Table 1.1. The self-developed servo drive is based on the TMS320F28335 Digital Signal Processor (DSP), which is used for

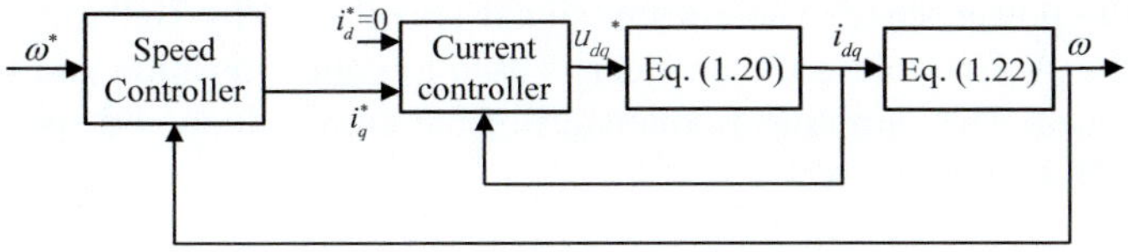

FIGURE 1.2 Block diagram of the dual-loop control of a PMSM.

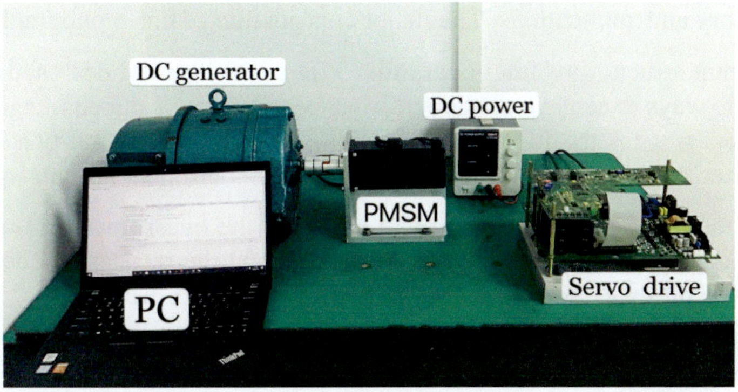

FIGURE 1.3 The PMSM speed control platform.

TABLE 1.1
Parameters of the Surface-Mounted Permanent Magnet Synchronous Motor (SPMSM)

Parameter	Unit	Value
Inertia J	kg.m^2	24.3
Pole Pairs p	-	5
Slot s	-	12
Permanent flux ψ	Wb	0.82
Stator resistance R	Ω	0.46
Stator inductance L	mH	2.4
Rated speed	rpm	2000
Rated torque	N·m	9.5
Rated current	A	9.4
Rated voltage	V	220
Rated power	W	2000

control algorithm implementation. The input command is given by the Ti Code Composer software (CCS) on PC and output data are sampled from encoder to servo drive by CCS. The insulated gate bipolar transistor (IGBT) power module PM100RLA060 is used as the power stage. The current signal is collected by Hall current sensors (LTSR 6-NP). A 23-bit absolute encoder is used to collect position and further calculate the speed. The sampling frequencies of the current loop and speed loop are 10k Hz and 2k Hz, respectively.

1.6 CONTRIBUTIONS

This book summarized the authors' research results on fractional order PID and ADR control theory and applications. The major contribution of this monograph include:

1. Different structure of fractional order PID controllers are designed in systematic ways to achieve better performances than the designed integer order controllers using traditional optimization. Stability and feasibility issues are analyzed and discussed for the fractional order controllers design.
2. The fractional order extended state observer, the quasi-fractional-resonant normalized extended state observer (NESO) and the resonant-model-based ESO are presented to deal with different types of disturbances.
3. Fractional order active disturbance rejection controllers, combining a fractional order PD controller and ESO/generalized ESO, are presented to improve dynamic tracking and disturbance rejection performance.
4. Simulation illustrations and/or experimental validations are demonstrated for all the proposed control schemes and approaches in this book.

1.7 ORGANIZATION

This book is organized as follows: The research motivation, fundamentals of fractional calculus and active disturbance rejection control, and monograph contributions are introduced in Part I; Part II presents different fractional order PID controllers, which include Chapters 2–5; Part III focuses on fractional order active disturbance rejection controls and includes Chapters 6–8; fractional order PID active disturbance rejection controls are presented in Part IV, including Chapters 9 and 10.

Part II

Fractional Order PID Control

2 Robust Three-Parameter FOPID Control

2.1 INTRODUCTION

The fractional order $PI^{\lambda}D^{\mu}$ controller [9], as introduced by Podlubny with its supplementary parameters of fractional order integral λ and fractional differential order μ, presents an extension of the classical proportional integral derivative (PID) controller. So far, various field-oriented controls (FOCs) have been proposed, demonstrating superior control efficacy over conventional integer order PID controllers [12, 15–17]. However, because of two more parameters of $PI^{\lambda}D^{\mu}$, its tuning method becomes much more difficult.

In the pursuit of optimal parameterization of the FOPID controller to enhance control performance, intelligent optimization algorithms such as differential evolution (DE) [18, 19], particle swarm optimization (PSO) [20], and chaotic firefly algorithms [21] have been efficaciously employed. These methodologies ascertain controller parameters by formulating and optimizing objective functions under constraints. However, these iterative processes often entail substantial temporal and computational resources, frequently yielding only locally optimal parameter sets.

The frequency-domain design approach for FOC is grounded in the frequency-domain characteristics of the control system [22] [23], whereby the control system is engineered to meet predefined frequency specifications such as gain crossover frequency ω_c, phase margin ϕ_m, and flat phase constraint. Nevertheless, this methodology cannot be straightforwardly applied to derive fractional order PID controllers due to the presence of five parameters. Thus, a three-parameter fractional order PID (TPFOPID) controller was introduced for processes with time-delay, demonstrating efficacy in various simulated and real-life applications [24, 25]. The proposed TP-FOPID incorporates three parameters by establishing interrelations between integral gain, differential gain, integral order, and differential order. This chapter aims to present a systematic design framework for TPFOPID controllers and delineate the attainable ranges of ω_c and ϕ_m for the PMSM speed servo system. Utilizing these predetermined achievable regions as foundational knowledge facilitates the determination of whether a robust controller can be engineered for a given combination of ω_c and ϕ_m for the PMSM speed servo system.

In this chapter, a TPFOPID controller design synthesis which can fulfill proper specifications – ω_c, ϕ_m, and a flat phase constraint – is proposed for the PMSM speed servo system. A two-dimensional graphics containing complete sets of feasible ω_c and ϕ can be drawn with this design scheme. Using this complete achievable region,

DOI: 10.1201/9781003397632-4

all the achievable combinations of ω_c and ϕ_m can be got before the controller design. Thus, only if choosing the combination of ω_c and ϕ_m from the achievable region can the desired TPFOPID controller be designed. Furthermore, the achievable region for TPFOPID is compared with this for the FOPI controller. The simulation illustration and experimental verification show that the proposed TPFOPID controller achieves better control performance over fractional order proportional integral (FOPI) and integer order proportional integral (IOPI) controllers, and the comparable control performance with the FOPID controller based on the DE complex optimal algorithm.

2.2 STABILIZING AND ROBUST TPFOPID CONTROLLER DESIGN

2.2.1 TPFOPID Controller

The considered feedback control structure is presented in Figure 2.1. r is the input reference, y is the output signal, and $C(s)$ is the designed TPFOPID, which is characterized with two coincident zeros z by the following transfer function [24],

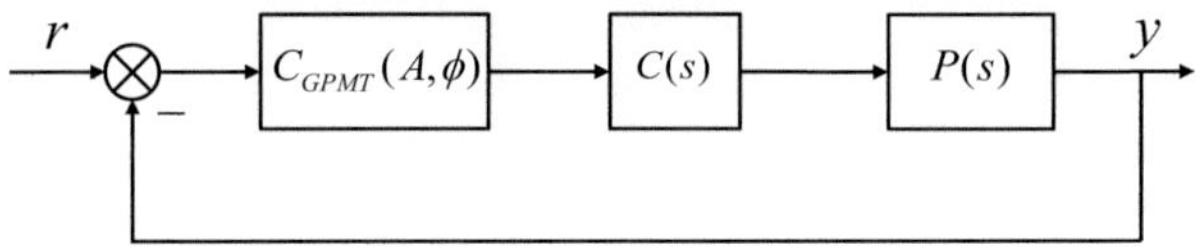

FIGURE 2.1 Feedback control system.

$$C(s) = K\frac{(s^{\lambda} - z)^2}{s^{\lambda}}, \tag{2.1}$$

with $K = K_p K_d$ and $z = \frac{-1}{2K_d}$, where K_p is the proportional gain, K_d is the derivative gain, and $\lambda \in (0,2)$ is the fractional order. The TPFOPID can also be expressed as

$$C(s) = K_p(1 + \frac{K_i}{s^{\lambda}} + K_d s^{\mu}), \quad K_i = \frac{1}{4K_d}, \quad \lambda = \mu. \tag{2.2}$$

where K_i is the integral gain, $\mu \in (0,2)$ are the fractional order. The reason for $K_i = \frac{1}{4K_d}$, i.e. two coincident zeros, is discussed in [26,27] for improving the tracking performance. The fractional orders λ and μ affect the magnitude Bode slope of the low and high frequencies, respectively, which can be clearly seen from the figure numbered 3 in [24]. The reason for choosing $\lambda = \mu$ is that the decrease in relative stability due to the integral is compensated by the increase in stability due to the derivative [24].

M_T, a gain-phase margin tester [28], is applied to plot boundaries of the TPFOPID controller with constant gain margin and phase margin in a parameter plane [29]. The transfer function is,

$$M_T(A,\phi) = Ae^{-j\phi}. \tag{2.3}$$

For finding the controller parameter boundaries for a given gain margin A of the control system, set $\phi = 0$ in (2.3). Given a prespecified $\phi = \phi_m$, the controller parameters boundaries can be plotted with $A = 1$.

$P(s)$ is the PMSM speed plant. Considering fractional characteristic and the common transfer function can be expressed as

$$P(s) = \frac{d}{s^\alpha + as^\beta + cs^\gamma + c}. \tag{2.4}$$

2.2.2 Stability Region Analysis

Because of the sensitivity of the differentiator to the high-frequency noise of the actual industrial control systems, a first-order low-pass filter is need to be introduced. Then the TPFOPID controller with the low-pass filter is described by (2.5),

$$C(s) = K_p(1 + \frac{1}{4K_d s^\lambda} + K_d s^\lambda)\frac{\omega_f}{s + \omega_f}, \tag{2.5}$$

where ω_f is the tuning frequency of the low-pass filter and $\omega_f = 80$ rad/s.

The closed-loop transfer function can be obtained as

$$T(s) = \frac{M_T(A,\phi)C(s)P(s)}{1 + M_T(A,\phi)C(s)P(s)}. \tag{2.6}$$

Substituting (2.4), (2.3), and (2.5) into (2.6) yields

$$T(s) = \frac{\omega_f dAK_p e^{-j\phi}(4K_d s^\lambda + 1 + 4K_d^2 s^{2\lambda})}{4K_d s^\lambda(s^\alpha + as^\beta + bs^\gamma + c)(s + \omega_f) + \omega_f dAK_p e^{-j\phi}(4K_d s^\lambda + 1 + 4K_d^2 s^{2\lambda})}. \tag{2.7}$$

Therefore, the characterized equation of the closed-loop system is

$$\begin{aligned} D(K_p, K_i, \lambda, A, \phi; s) = {} & 4K_d s^\lambda(s^\alpha + as^\beta + bs^\gamma + c)(s + \omega_f) \\ & + \omega_f dAK_p e^{-j\phi}(4K_d s^\lambda + 1 + 4K_d^2 s^{2\lambda}). \end{aligned}$$

The root locations of the characterized equation determines the stability of the system (2.8) with $\phi = 0°$ and $A = 1$. The boundaries of the controller parameters can be determined by fractional order λ, real root boundary (RRB), infinite root boundary (IRB) and complex root boundary (CRB) [29–31]:

i. Region of λ: The fractional order λ in the TPFOPID controller belongs to $(0,2)$.
ii. RRB: The RRB can be got by the equation $D(K_p, K_d, \lambda, A, \phi; s = 0)$, that is $K_p = 0$.

iii. IRB: The IRB can be descried by the following equations:

$$K_d = \begin{cases} 0, & \text{for } (\lambda > \alpha) \\ \pm\frac{1}{d}, & \text{for } (\lambda = \alpha) \\ none, & \text{for } (\lambda < \alpha) \end{cases} \tag{2.8}$$

iv. CRB: Substituting $s = j\omega$ into (2.8) and considering the real part and the imaginary part, respectively,

$$4K_d(E_1\omega_f - \omega E_2) + \omega dAK_p(C_7E_3 + S_7E_4) = 0, \tag{2.9}$$

$$4K_d(\omega E_1 + \omega_f E_2) + \omega_f dAK_p(C_7E_4 - S_7E_3) = 0, \tag{2.10}$$

where

$$E_1 = \omega^{\lambda+\alpha}C_1 + a\omega^{\lambda+\beta}C_2 + b\omega^{\lambda+\gamma}C_3 + \omega^{\lambda}C_4,$$
$$E_2 = \omega^{\lambda+\alpha}S_1 + a\omega^{\lambda+\beta}S_2 + b\omega^{\lambda+\gamma}S_3 + \omega^{\lambda}S_4,$$
$$C_1 = \cos\frac{(\lambda+\alpha)\pi}{2}, \quad C_2 = \cos\frac{(\lambda+\beta)\pi}{2}, \quad C_3 = \cos\frac{(\lambda+\gamma)\pi}{2}, \quad C_4 = \cos\frac{\lambda\pi}{2},$$
$$S_1 = \sin\frac{(\lambda+\alpha)\pi}{2}, \quad S_2 = \sin\frac{(\lambda+\beta)\pi}{2}, \quad S_3 = \sin\frac{(\lambda+\gamma)\pi}{2}, \quad S_4 = \sin\frac{\lambda\pi}{2},$$
$$C_7 = \cos(\phi), \quad S_7 = \sin(\phi).$$

From (2.9) and (2.10),

$$K_d = \frac{-B \pm \sqrt{B^2 - 4A_1C}}{2A_1}, \tag{2.11}$$

$$K_p = \frac{4K_d(\omega E_2 - E_1\omega_f)}{\omega_f dA(C_7E_3 + S_7E_4)}, \tag{2.12}$$

where,

$$A_1 = 4\omega^{2\lambda}C_6(C_7 + S_7F) + 4\omega^{2\lambda}S_6(S_7 - C_7F),$$
$$B = 4\omega^{\lambda}C_5(C_7 + S_7F) + 4\omega^{\lambda}S_5(S_7 - C_7F), \quad C = C_7 + S_7F,$$
$$C_5 = \cos(\frac{\lambda\pi}{2}), \quad C_6 = \cos(\lambda\pi), \quad S_5 = \sin(\frac{\lambda\pi}{2}), \quad S_6 = \sin(\lambda\pi),$$
$$F = \frac{E_1\omega_f - \omega E_2}{\omega E_1 + \omega_f E_2}.$$

Fixed λ, the curve of K_d vs. K_p can be plotted with ω from $0 \to +\infty$.

Therefore, given $A = 1$, $\phi = 0°$ and an arbitrary fractional order $\lambda_1 = 0.5$, the line of K_p w.r.t K_d can be plotted according to the Eqs. (2.11) and (2.12) of the CRB and the line $K_p = 0$ of the RRB as shown (black line) in Figure 2.2a. The stability region can be determined by a random point testing [32]. Sweeping all the $\lambda \in (0,2)$, the complete stability region for the TPFOPID controller can be obtained as shown in Figure 2.2b.

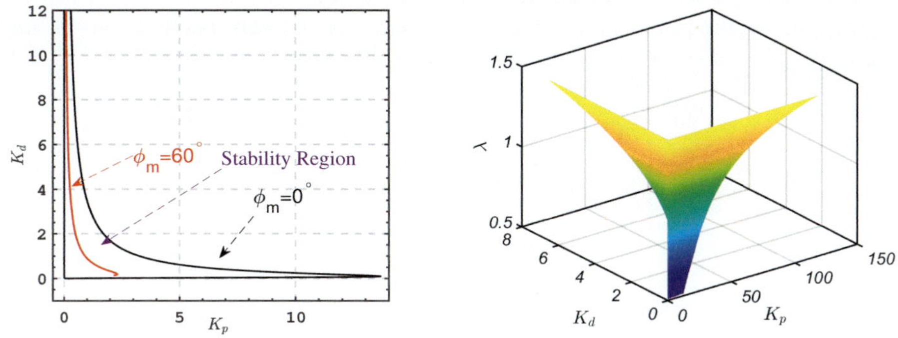

(a) Stability region with $\phi_m = 0°$ as black line and $\phi_m = 60°$ as red line ($\lambda = 0.5$).

(b) Complete stability surface for $\phi_m = 0°$.

FIGURE 2.2 Stability boundary comparison of K_p vs. K_d and three dimensional complete stability surface.

2.2.3 Design Scheme

Setting $A = 1$ and $\phi_m = 60°$, the *relative stability line* can be plotted as shown (red line) in Figure 2.2a in the (K_p, K_d)-plane with a fixed $\lambda_1 = 0.5$. Note, the parameters, points on this boundary are related to the gain crossover frequency [33]. Given another specification gain crossover frequency $\omega_c = 80$ rad/s, a point (K_p, K_d) on the *relative stability line* can be determined with λ_1 as shown in Figure 2.3.

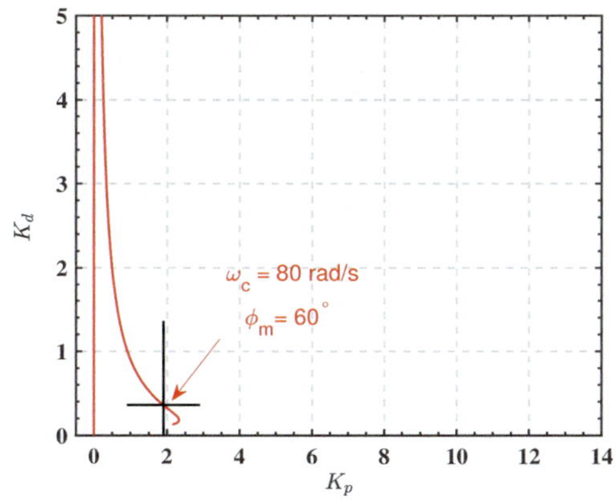

FIGURE 2.3 The designed K_p and K_d satisfying $\omega = 80$ rad/s, $\phi_m = 60°$ with $\lambda =$ 0.5.

Therefore, with the prespecified ω_c, ϕ_m and a fixed λ_1, the other two parameters K_p and K_d of TPFOPID can be fixed on the *relative stability lines*. A curve named *relative stability curve* can be determined by sweeping all the λ in $(0,2)$ as shown in Figure 2.4. All the points on this curve can guarantee the two specifications ω_c and ϕ_m.

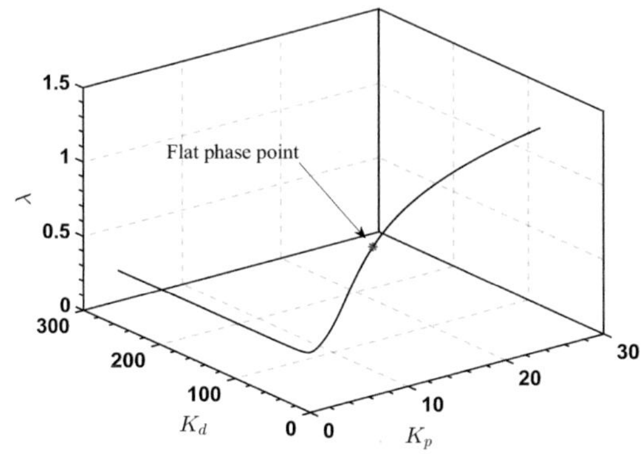

FIGURE 2.4 The flat phase stable point on the relative stability curve.

The flat phase means the phase is equal around the ω_c point in the open-loop Bode plot, which can make the system robust to the system gain variations [34]. That is, the overshoots of the step response are substantially equal despite the open-loop gain variation in a certain range.

From (2.9) and (2.10),

$$S_7 = \frac{B_2E_3 - B_1E_4}{AK(E_3^2+E_4^2)}, \tag{2.13}$$

$$C_7 = \frac{-B_1E_3 - B_2E_4}{AK(E_3^2+E_4^2)}, \tag{2.14}$$

where

$$B_1 = 4K_d(E_1\omega_f - \omega E_2), \quad B_2 = 4K_d(\omega E_1 + \omega_f E_2), \quad K = d\omega_f K_p.$$

From (2.13) and (2.14),

$$\phi = \arctan\frac{B_1E_4 - B_2E_3}{B_1E_3 + B_2E_4} + n\pi. \tag{2.15}$$

To satisfy the flat phase constraint, keep that

$$\frac{\mathrm{d}\theta}{\mathrm{d}\omega} = 0.$$

According to the above analysis, $\theta = \phi - \pi$. Thus,

$$\frac{\mathrm{d}\theta}{\mathrm{d}\theta} = \frac{\mathrm{d}\phi}{\mathrm{d}\omega} = 0.$$

From (2.15),

$$\frac{d\phi}{d\omega} = \frac{(B_1^2+B_2^2)(E_3E_4'-E_3'E_4)+(B_1'B_2-B_1B_2')(E_3^2+E_4^2)}{(B_1E_3+B2E_4)^2+(B_1E_4-B_2E_3)^2} = 0, \tag{2.16}$$

where,

$$E_3' = 4K_d\lambda\omega^{\lambda-1}C_5+8K_d^2\lambda\omega^{2\lambda-1}C_6, \quad E_4' = 4K_d\lambda\omega^{\lambda-1}S_5+8K_d^2\lambda\omega^{2\lambda-1}S_6,$$
$$B_1' = 4K_d(\omega_f E_1'-(E_2+\omega E_2')), \quad B_2' = 4K_d(E_1+\omega E_1'+\omega_f E_2'),$$
$$E_1 = (\lambda+\alpha)\omega^{\lambda+\alpha-1}C_1+a(\lambda+\beta)\omega^{\lambda+\beta-1}C_2+b(\lambda+\gamma)\omega^{\lambda+\gamma-1}C_3+\lambda\omega^{\lambda-1}C_4,$$
$$E_2 = (\lambda+\alpha)\omega^{\lambda+\alpha-1}S_1+a(\lambda+\beta)\omega^{\lambda+\beta-1}S_2+b(\lambda+\gamma)\omega^{\lambda+\gamma-1}S_3+\lambda\omega^{\lambda-1}S_4.$$

All points on the *relative stability curve* satisfying the specifications ω_c and ϕ_m can be tested through Eq. (2.16). The *flat phase stable point* satisfying three specifications will be found as a red star as shown in Figure 2.4. Thus, the goal of designing a robust TPFOPID controller for the PMSM speed-loop system have been achieved.

2.2.4 Achievable Region Collection

From Sec. 2.2.3, the desired TPFOPID can be obtained by giving a fixed phase margin ϕ_m and a certain gain crossover frequency ω_c. Then, with the fixed phase margin $\phi_m = 60°$, the region of gain crossover frequency can be obtained as ω_c increasing from 0 as shown in Figure 2.5. Then, sweeping all the phase margin from 0 to 2π, the complete achievable region for the phase margin and gain crossover frequency while

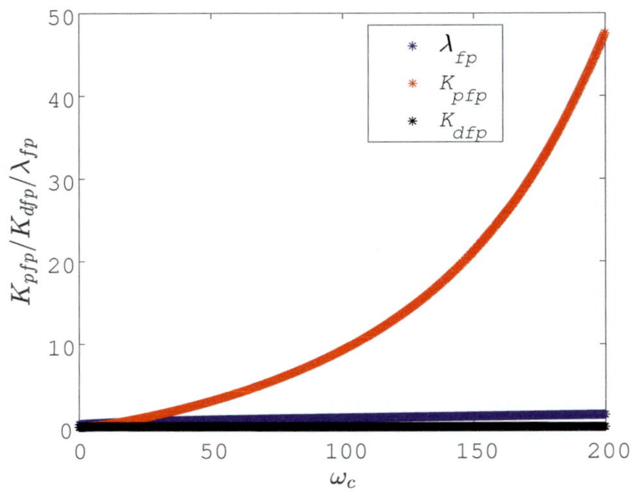

FIGURE 2.5 K_{pfp}, K_{dfp}, and λ_{fp} vs. ω_c with $\phi_m = 60°$.

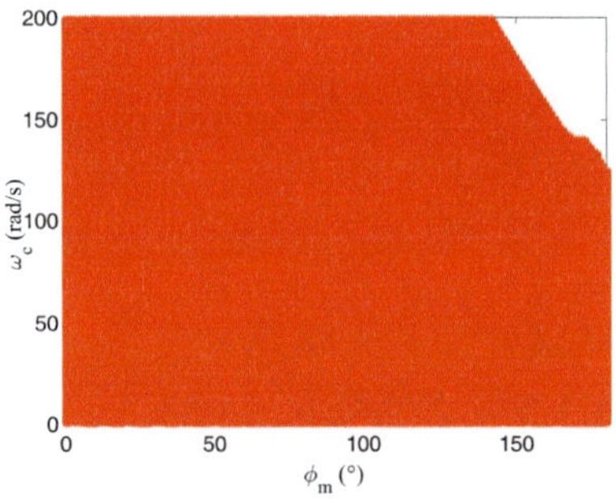

(**a**) TPFOPID controller.

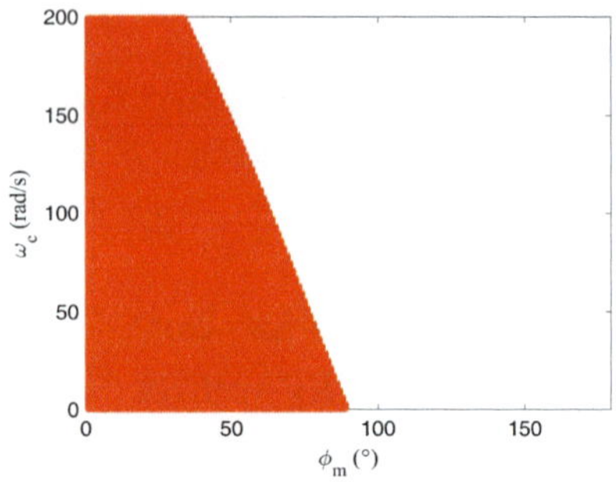

(**b**) FOPI controller.

FIGURE 2.6 Achievable region of TPFOPID and FOPI.

fulfill the flat phase, can be collected as show in Figure 2.6a. With the information of the complete achievable region, users can choose the combination of ω_c and ϕ_m from Figure 2.6a before controller design.

2.3 ACHIEVABLE REGION FOR FOPI CONTROLLER

In this section, the FOPI is also designed with specifications ω_c, ϕ_m, and the flat phase constraint for PMSM speed servo system. The complete achievable regions for FOPI are shown in Figure 2.6b.

According to Figure 2.6a and 2.6b, the achievable region of ϕ_m and ω_c for the proposed TPFOPID controller is significantly bigger than that for FOPI controller. That means the proposed TPFOPID controller give user more options to design the controller which satisfy desired control performance.

2.4 EXPERIMENTAL STUDY

In motor control experiments, the fractional order operator s^r can be implemented by the impulse response invariant discretization (IRID) method. For example, the approximate finite dimensional discretized (z) transfer function of $s^{-0.9877}$ has the following form,

$$s^{-0.9877} \approx \frac{Num}{Den}, \tag{2.17}$$

where

$$Num = 0.0005998z^5 - 0.00135z^4 + 0.0009605z^3 - 0.0001908z^2 - 2.254e-05z + 4.753e-06,$$

$$Den = z^5 - 3.383z^4 + 4.31z^3 - 2.519z^2 + 0.6415z - 0.05011,$$

and the comparison of the approximated Bode diagram and true Bode diagram is shown in Figure 2.7.

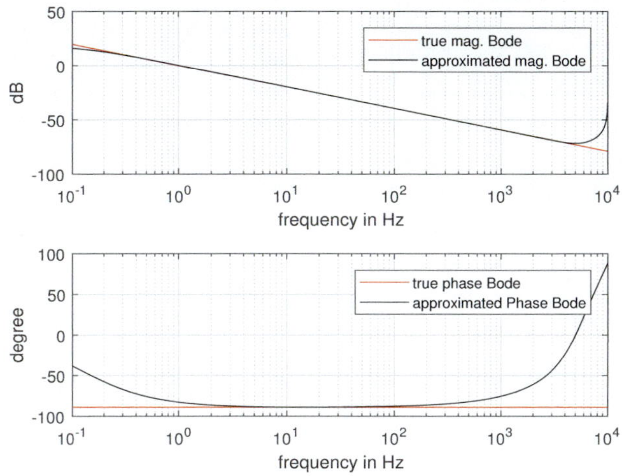

FIGURE 2.7 Comparison of the approximated Bode diagram and true Bode diagram.

2.4.1 Example 1: Robust Verification

Setting the reference speed as 600 r/min, PMSM speed response experiments are performed, using the designed TPFOPID and the IOPI controllers to control the motor speed. The step responses of two control systems are shown in Figure 2.8. Setting the IOPI and TPFOPID controller gains as 80%, 100%, and 120% of the nominal

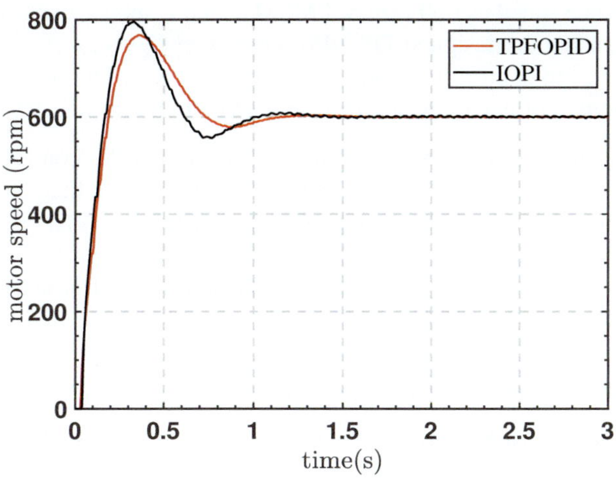

FIGURE 2.8 The speed responses of IOPI and TPFOPID control systems (experiment).

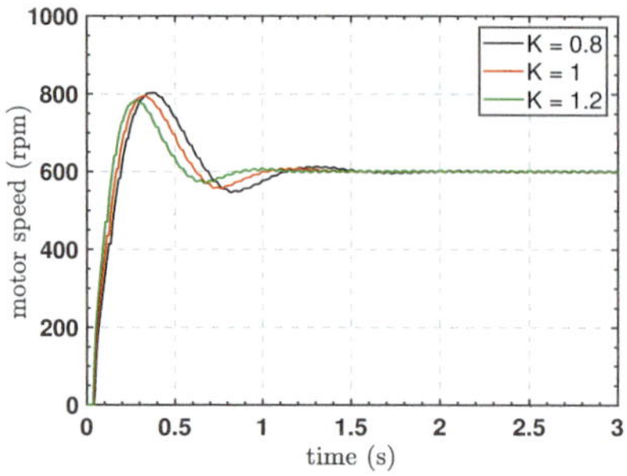

(**a**) IOPI control system.

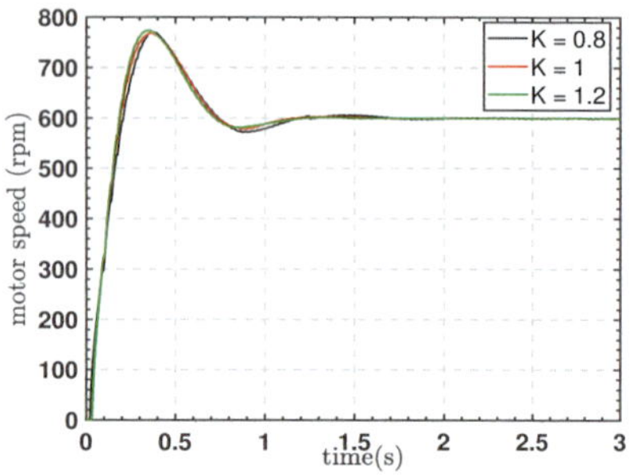

(**b**) TPFOPID control system.

FIGURE 2.9 Robustness comparison of IOPI and TPFOPID control systems (experiment).

value, the motor speed tracking experiments are carried out with the reference speed as 600 r/min. The experimental results are shown in Figure 2.9a and 2.9b.

According to Figure 2.8, the designed TPFOPID control system has less overshoot than IOPI control system. In terms of robustness to the system loop gain variations, the overshoots of TPFOPID control system are almost equal but those of the IOPI control system are obviously different. Thus, the proposed TPFOPID controller achieves better tracking performance and robustness to system loop gain variations than traditional the IOPI controller.

2.4.2 Example 2: Comparison with FOPI

Setting the reference speed as 600 r/min, PMSM speed response experiment results are performance, using the designed TPFOPID controller $C_{TPFOPID1}(s)$ and the FOPI controller $C_{FOPI1}(s)$ to control the motor speed, respectively. The step and anti-load disturbance responses of two control systems are shown in Figure 2.10.

According to Figure 2.10, the speed response of the TPFOPID control system has smaller overshoot (37.83% and 45.33%) and smaller oscillation than the FOPI control system. Therefore, the TPFOPID control system achieves better speed response performance. Besides, the response of the FOPI control system has bigger speed drop and longer recovery time. Therefore, the TPFOPID control system achieves better anti-load disturbance performance.

2.4.3 Example 3: Achievable Region Verification

Setting the reference speed as 600 r/min, motor speed-tracking and anti-load disturbance experiments are performed, using the designed TPFOPID controller $C_{TPFOPID2}(s)$ and FOPI controller $C_{FOPI2}(s)$. The responses are shown in Figure 2.11.

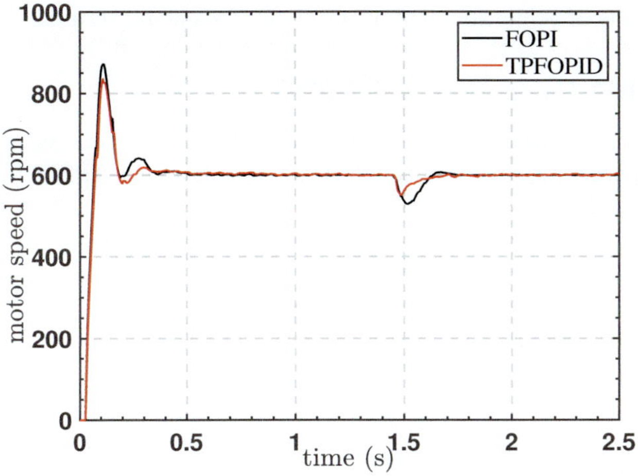

FIGURE 2.10 The speed responses and anti-load disturbance responses of the TPFOPID and FOPI control systems (experiment).

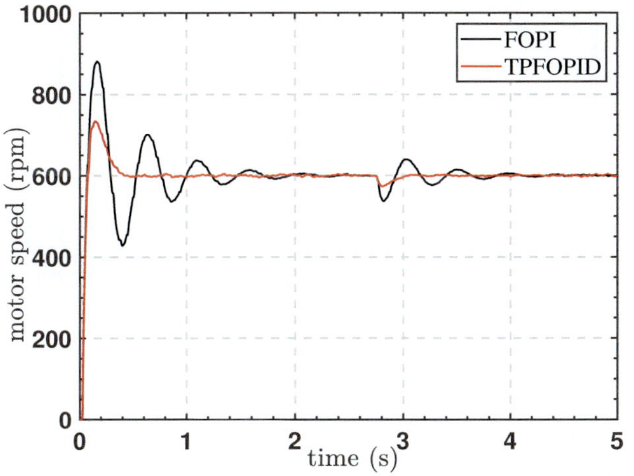

FIGURE 2.11 The speed responses and anti-load disturbance responses of the TPFOPID and FOPI control systems (experiment).

According to Figure 2.11, the response of the TPFOPID control system has smaller overshoot and shorter settling time. Moreover, the response of the FOPI control system has obvious oscillation. Therefore, the TPFOPID control system achieves better step response performance. Furthermore, the anti-load disturbance response of the TPFOPID control system has smaller speed drop and the FOPI control system has obvious oscillation and longer recovery time. Therefore, the TPFOPID control system achieves better anti-load disturbance performance. Moreover, the comparison results of achievable regions in Figure 2.6 can be verified.

2.4.4 Example 4: Comparison with FOPID

Setting the reference speed as 600 r/min, the motor speed-tracking and anti-load disturbance experiments are performed, using the designed TPFOPID controller $C_{TPFOPID3}(s)$ and FOPID controller $C_{FOPID}(s)$. The responses of two control systems are shown in Figure 2.12. Two control systems achieve similar rising time (0.072 s and 0.073 s) and overshoot (19.6% and 19.3%). And the speed drops of the two control systems are 3.73% and 3.83%, respectively. The comparison results are summarized in Table 2.1.

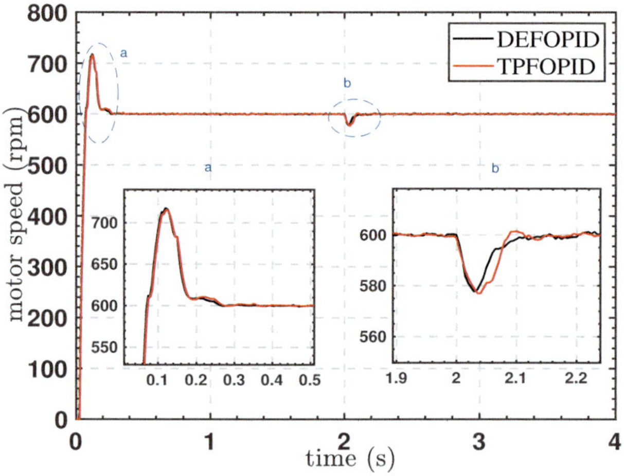

FIGURE 2.12 The speed responses and anti-load disturbance responses of the TP-FOPID and FOPID control systems (experiment).

TABLE 2.1
Comparison of Responses Results with Two Control Systems (Experiment)

Controller	Rising Time(s)	Overshoot(%)	Speed Drop(%)
TPFOPID	0.073	19.3	3.88
FOPID	0.072	19.6	3.73

Obviously, two control systems achieve extremely close speed-tracking and anti-load disturbance performances, although the TPFOPID controller only has three parameters and the tuning method is greatly simplified, which makes the FOPID controller more acceptable to practical industrial applications.

2.5 CHAPTER SUMMARY

In this chapter, a TPFOPID controller design method is proposed for the PMSM speed servo system, by establishing the relationship between the integral gain K_i, the derivative gain K_d, and the fractional orders λ and μ. Moreover, a synthesis scheme to design the desired TPFOPID controller satisfying prespecified ω_c, ϕ_m and the flat phase constraint for the PMSM speed servo system is proposed. This designed TPFOPID controller can get desired control performance which is robust to the loop gain variations. Beyond that, the entire feasible region of two specifications can be collected for the proposed TPFOPID controller for the PMSM speed servo system. With this region as priori information, all the feasible combinations of ω_c and ϕ_m can be determined before robust controller design. The feasible region of the proposed TPFOPTD controller is bigger than FOPI controller, which also has three parameters. This indicates that the TPFOPID controller, which incorporates all three components (proportional, integral, and derivative) yet requires only three parameters, can better satisfy users' comprehensive requirements for both response speed and relative stability in control systems compared to FOPI controllers. Simulation and experimental results show that the TPFOPID controller achieves better step response performance and better robustness to system gain variations than IOPI and FOPI controllers designed by the same method. Moreover, the designed TPFOPID controller is also compared with the FOPID controller with its complex tuning method. The results show that the TPFOPID controller with fewer parameters and simpler tuning method presents comparable control performance to that of the FOPID controller.

3 Optimal Four-Parameter $FOPI^{\lambda}D$ Control

3.1 INTRODUCTION

Many industrial processes can be described or fitted by the first order plus time delay (FOPTD) plant [35, 36]. It has attracted much scholarly attention over the years. [37, 38]. The proportional integral derivative (PID) controller plays an essential role in the control engineering industry in the past few decades [39]. Despite its simple structure, the PID controller has proven effective in most control problems. In recent years, fractional calculation has attracted great attention of many scholars, especially in the fields of physics, biology, and engineering [40–42]. The FO-$PI^{\lambda}D^{\mu}$ controller was proposed by Podlubny [9], demonstrating it can achieve better control performance compared to the classical PID controller because of the additional two parameters λ and μ. Until now, various types FOPID controllers were proposed for integer order (IO) and fractional order (FO) plants [5, 12, 15, 43, 44].

In fact, the tuning method for fractional order controller (FOC) is always a hot topic and also difficult because of additional parameters fractional orders λ and μ. So far, the popular methods for tuning FOC are frequency-domain analytical methods and time-domain optimization methods. For instance, some frequency-domain analytical methods obtain the controller parameters by solving a set of equations representing specifications related to phase margin, ω_{gc}, flat phase constraint, sensitivity function, and so on [1, 45–47]. In [47], an interesting structure, a FOPID cascaded with a fractional filter (FF), was also proposed for higher-order processes based on Bode's ideal. However, the frequency-domain methods are not able to guarantee the optimal dynamic control performance. The time-domain optimization methods obtain FOC parameters by utilizing optimization algorithms to minimize certain performance indices [48–50], e.g., the integral time absolute error (ITAE), the integral square error (ISE), etc., which ensure the optimal dynamic control performance of the systems. However, these time-domain methods are not able to guarantee some desired frequency-domain specifications.

As an extension of Chapter 2, this chapter proposes a tuning method of the $FOPI^{\lambda}D$ controller for all the FOPTD systems, which can satisfy both the frequency-domain specifications (i.e. , ω_{gc}, phase margin and flat phase constraint) and time-domain optimal performance requirement (i.e. ITAE criterion). Thus, the designed control system not only meets the given frequency-domain specifications, but also achieves the optimal dynamic control performance. Moreover, a synthesis scheme is proposed to obtain a complete achievable region of two specifications on ω_{gc} and a phase

DOI: 10.1201/9781003397632-5

margin of robust FOPI$^\lambda$D controller. User can choose the specification combination from the achievable region before the FOPI$^\lambda$D controller design. The simulation results demonstrated that the proposed FOPI$^\lambda$D is superior to other FOCs in references [24, 33, 47, 51, 52].

3.2 STABILITY REGION ANALYSIS

Figure 3.1 is the considered control system; r and y are the input and output actual signals, respectively; and e, u, d are the error signal, controller output, and disturbance input, respectively. $C(s)$ is the proposed FOPI$^\lambda$D controller as below:

$$C(s) = K_p + \frac{K_i}{s^\lambda} + K_d s, \tag{3.1}$$

where K_p, K_i, K_d are proportional, integral, and derivative gains, respectively. $\lambda \in (0,2)$ is fractional order.

$P(s)$ is the FOPTD system as below:

$$P(s) = \frac{K}{Ts+1}e^{-Ls}, \tag{3.2}$$

where K, T, and L are the plant gain, time constant, and pure delay, respectively. Many industrial processes can be described using an FOPTD model.

The gain-phase margin tester (2.3) is applied to plot the parameter boundaries of the controller with the phase margin (ϕ) and gain margin (A). In this chapter, A will be set as 1 to meet the given ϕ_m.

The closed-loop transfer function in Figure 3.1 is

$$G(s) = \frac{C(s)M_T(A,\phi)P(s)}{1+C(s)M_T(A,\phi)P(s)}. \tag{3.3}$$

The denominator of (3.3) is the characteristic equation of the control system. Substituting (3.1) (3.2) (2.3) into (3.3), yields,

$$D(s;K_p,K_i,\lambda,K_d,A,\phi) = s^\lambda(Ts+1) + (K_p s^\lambda + K_i + K_d s^{1+\lambda})Ae^{-j\phi}e^{-Ls}. \tag{3.4}$$

The stability region boundaries can be illustrated by real root boundary (RRB), infinite root boundary (IRB) and complex root boundary (CRB) [31, 33, 53]:

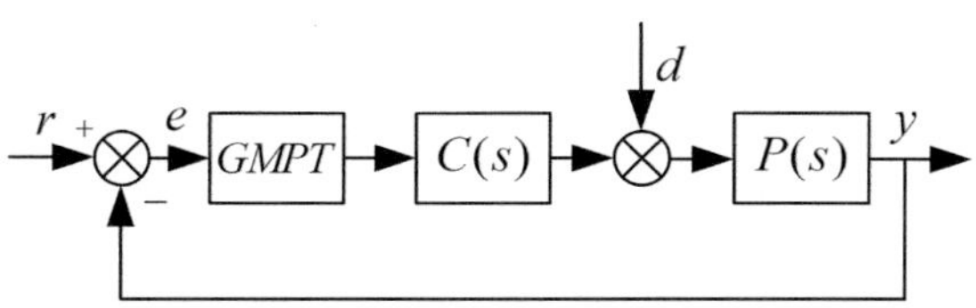

FIGURE 3.1 Feedback control system.

i. Reachable interval of λ: The interval of λ is $\lambda \in (0,2)$.

ii. RRB: The RRB is described as $D(s=0;K_p,K_i,\lambda,K_d,A,\phi)=0$, so the boundary of K_i is,

$$K_i = 0.$$

iii. IRB: Due to the relative order of $P(s)$ is 1, the IRB is decided by the equation $D(s=+\infty;K_p,K_i,\lambda,K_d,A,\phi)=0$, and the result is [29],

$$K_d = \pm\frac{T}{K}. \tag{3.5}$$

iv. CRB: putting $j\omega$ for s into (3.4), the CRB can be defined from $D(s=j\omega;K_p,K_i,\lambda,K_d,A,\phi)$ as follows,

$$\begin{aligned} &D(j\omega;K_p,K_i,\lambda,K_d,A,\phi)= \\ &(j\omega)^{\lambda}(jT\omega+1)+(K_p(j\omega)^{\lambda}+K_i+K_d(j\omega)^{1+\lambda})Ae^{-j\phi}e^{-jL\omega}=0. \end{aligned} \tag{3.6}$$

Then, the real and imaginary items of (3.6) equal zero,

$$B_1+AKC_1E+AKS_1F=0, \tag{3.7}$$
$$B_2+AKC_1F-AKS_1E=0, \tag{3.8}$$

where

$$\begin{aligned} &B_1=\omega^{\lambda}C_2-T\omega^{1+\lambda}S_2, \quad B_2=T\omega^{\lambda+1}C_2+\omega^{\lambda}S_2, \\ &E=K_p\omega^{\lambda}C_2+K_i+K_d\omega^{1+\lambda}C_3, \quad F=K_p\omega^{\lambda}S_2+K_d\omega^{1+\lambda}S_3, \\ &C_1=\cos(\phi+\omega \mathrm{L}), \quad C_2=\cos\frac{\lambda\pi}{2}, \quad C_3=\cos\frac{(1+\lambda)\pi}{2}, \\ &S_1=\sin(\phi+\omega \mathrm{L}), \quad S_2=\sin\frac{\lambda\pi}{2}, \quad S_3=\sin\frac{(1+\lambda)\pi}{2}. \end{aligned}$$

Solving the equation set (3.7) and (3.8), the K_p and K_i represented by λ and K_d are shown as

$$K_p=-\frac{B_1S_1+B_2C_1}{AK\omega^{\lambda}S_2}-\frac{K_d\omega S_3}{S_2}, \tag{3.9}$$

$$K_i=\frac{B_2-B_1C_1S_1-B_2C_2^2}{AKS_1}+\frac{B_1S_1C_2+B_2C_1C_2}{AKS_2}+\frac{K_d\omega^{1+\lambda}S_3C_2}{S_2}-K_d\omega^{1+\lambda}C_3. \tag{3.10}$$

Fixed K_d and λ, the above two equations plot a line of K_p versus K_i with ω from $0 \to +\infty$.

Thus, for $A=1$ and $\phi=0$, the IRB, CRB, and RRB with the fixed values of λ and K_d are determined. Thus, the parameter space is classified as unstable and stability regions. Random-point testing [32] can find the stability region.

3.3 THE OPTIMAL ROBUST FOPI$^\lambda$D CONTROLLER DESIGN

3.3.1 Controller Design Specifications

In this chapter, three frequency-domain specifications phase margin (ϕ_m), ω_{gc}, and flat phase constraint are applied to the design FOPI$^\lambda$D controller:

- Gain crossover frequency

$$|C(j\omega_{gc})P(j\omega_{gc})|_{dB} = 0. \tag{3.11}$$

- Phase margin

$$\arg[C(j\omega_{gc})P(j\omega_{gc})] = -\pi + \phi_m. \tag{3.12}$$

- Flat phase constraint (The phase Bode plot is flat at ω_{gc}, indicating that the system is robust to variations in loop gain. In other words, it maintains a consistent overshoot despite changes in gain.)

$$\frac{d\phi_m}{d\omega_{gc}} = 0. \tag{3.13}$$

One more design constraint is needed because there are four controller parameters. A time-domain specification, ITAE, is applied to realize an optimal dynamic performance:

- The cost function, i.e., ITAE:

$$J = \int_0^\infty t|e(t)|\,dt, \tag{3.14}$$

where $e(t)$ is the deviation between the input reference signal and the actual output signal.

3.3.2 Design with Frequency-Domain Specifications

From Sec. 3.3.1, the stability region can be determined with fixed K_d and λ. Then the line in the stability region can be obtained to meet the given ϕ_m with $A = 1$ in (3.6).

Consider one frequency-domain specification: phase margin ϕ_m. The *relative stable line* (RSL) will be obtained in the parameter space as $\omega \to \omega_1$ from zero with a certain K_d and λ for $A = 1$ and $\phi = \phi_m$ in (3.6). ω_1 is the maximum frequency in the stability region with the certain K_d and λ, guaranteeing the prespecified phase margin.

Note that the points on RSL are related to the ω_{gc} [33]. Thus, with the prespecified ϕ_m, ω_{gc}, a certain K_d and λ and the other two FOPI$^\lambda$D parameters K_p and K_i can be fixed to be a point on the above RSL with $\omega = \omega_{gc}$ and $\phi = \phi_m$. Sweeping over the

λ in $(0,2)$, a curve named the *relative stability curve* (RSC) can be obtained, and all the points on RSC satisfy the specifications ω_{gc} and ϕ_m.

The flat phase constraint is applied to design the FOPI$^\lambda$D. From (3.7) and (3.8),

$$S_1 = \frac{B_2E - B_1F}{AK(E^2+F^2)}, \tag{3.15}$$

$$C_1 = \frac{-B_1E - B_2F}{AK(E^2+F^2)}. \tag{3.16}$$

So, one can get

$$\phi = \arctan\frac{B_1F - B_2E}{B_1E + B_2F} - \omega L + n\pi. \tag{3.17}$$

In order to satisfy that the phase Bode plot is flat at ω_{gc}, we have

$$\frac{\mathrm{d}\phi}{\mathrm{d}\omega} = \frac{(B_1^2+B_2^2)(EF' - E'F) + (B_1'B_2 - B_1B_2')(E^2+F^2)}{(B_1E+B_2F)^2 + (B_1F - B_2E)^2} - L = 0, \tag{3.18}$$

where

$$E' = \lambda K_p\omega^{\lambda-1}C_2 + (1+\lambda)K_d\omega^\lambda C_3, \quad F' = \lambda K_p\omega^{\lambda-1}S_2 + (1+\lambda)K_d\omega^\lambda S_3,$$
$$B_1' = \lambda\omega^{\lambda-1}C_2 - (1+\lambda)T\omega^\lambda S_2, \quad B_2' = (\lambda+1)T\omega^\lambda C_2 + \lambda\omega^{\lambda-1}S_2.$$

So, all the points on the RSC corresponding to the FOPI$^\lambda$D parameters can be tested by Eq. (3.18). One specific point can be obtained which satisfies the relationship (3.18). Sweeping all the K_d in $[-\frac{T}{K}, +\frac{T}{K}]$, one curve named *flat phase curve* (flat phase constraint) can be obtained. All the points on flat phase constraint can ensure the three specifications ω_{gc}, ϕ_m, and flat phase.

3.3.3 Design with Time-Domain Specifications

Given all the parameter sets of the FOPI$^\lambda$D corresponding to the points on the flat phase curve, the step responses in simulation are implemented and the ITAEs are calculated. Eq. (3.19) is applied for ITAE calculation,

$$J_{ITAE} = \sum_{n=1}^{\frac{T}{\Delta t}} t[n]|e[n]|\Delta t, \tag{3.19}$$

where T is simulation time, and Δt is sampling time. Eq. (3.19) is the discrete form of Eq. (3.14).

Thus, the relationship plot between λ and J_{ITAE} can be obtained. Choosing the smallest J_{ITAE}, the parameters (K_p, K_i, λ, K_d) of FOPI$^\lambda$D can be determined, which is the optimal robust FOPI$^\lambda$D controller satisfying both frequency-domain and time-domain specifications simultaneously.

3.3.4 Achievable Region Collection

According to Sec. 3.3.2, the ω_1 can be determined with the given phase margin, fixed λ and K_d. Sweeping all the $\lambda \in (0,2)$, the upper boundary ω_{1max} will be found [33]. Thus, searching the frequencies in $(0, \omega_{1max}]$ with a fixed ϕ_m, the region of ω_{gc} can be decided. Then, sweeping all the ϕ_m in $(0, 2\pi)$, the achievable region of ω_{gc} and ϕ_m can be decided. Afterwards, sweeping all the $K_d \in [-\frac{T}{K}, +\frac{T}{K}]$, the entire achievable region of ω_{gc} and ϕ_m, guaranteeing the flat phase constraint, can be collected.

3.3.5 Design Guideline Summary

In this subsection, the design guidelines for optimal robust FOPI$^{\lambda}$D controller are summarized in Figure 3.2.

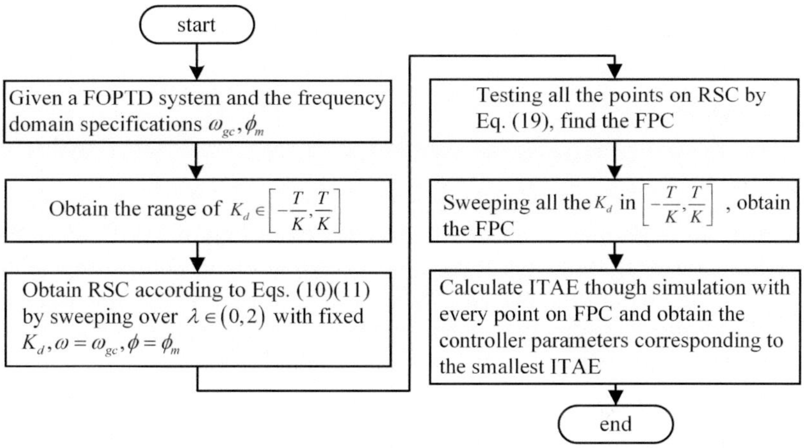

FIGURE 3.2 The design guidelines summary.

3.4 DESIGN SCHEME SUMMARY WITH AN EXAMPLE

According to the discussion above, the design scheme summary with an example is presented in this section.

Step 1. Given a FOPTD system with $L = 0.1$, $T = 1$, $K = 1$ and with prespecified $\omega_{gc} = 5$ rad/s and phase margin $\phi_m = 50°$.

Step 2. According to the IRB, the stability interval of K_d is $[-1, 1]$ according to the Eq. (3.5). Choosing $K_d = 0$ and $\lambda = 0.5$, the CRB line of K_p w.r.t. K_i and RRB line is obtained in the (K_p, K_i) space as red line in Figure 3.3 with $\phi = 0°$. Through random-point-testing in each region, the stability and unstable region can be determined.

Step 3. Given the $\phi = \phi_m$, with fixed $K_d = 0$ and $\lambda = 0.5$, the RSL is plotted as shown with the black line in Figure 3.3. All the points on this line satisfy the phase margin $\phi_m = 50°$.

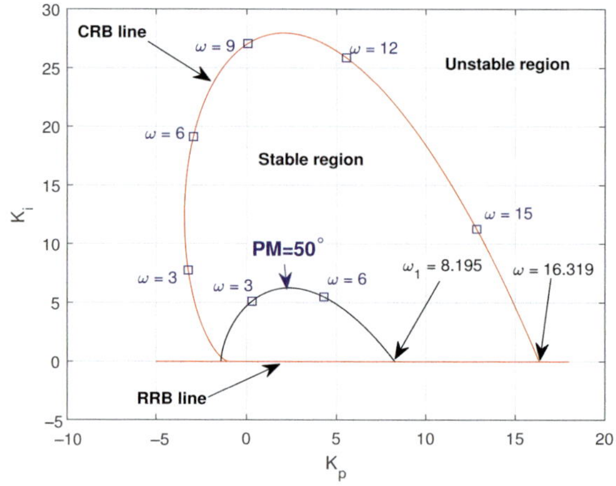

FIGURE 3.3 Stability region of K_p and K_i with $\phi_m = 0^\circ$ as red line and $\phi_m = 50^\circ$ as black line ($\lambda = 0.5$, $K_d = 0$).

Step 4. Given the prespecified $\omega_{gc} = 5$ rad/s, the parameters K_p and K_i can be determined as the blue triangle on the RSL as shown in Figure 3.4. Given different λ, the RSL and corresponding K_p and K_i are shown in Figure 3.4. Sweeping all the $\lambda \in (0,2)$, all the parameter sets (K_p, K_i, λ) can be plotted as the blue triangles in Figure 3.5. All the controller parameter sets corresponding to the blue triangles can meet the two specifications $\phi_m = 50^\circ$ and $\omega_{gc} = 5$ rad/s.

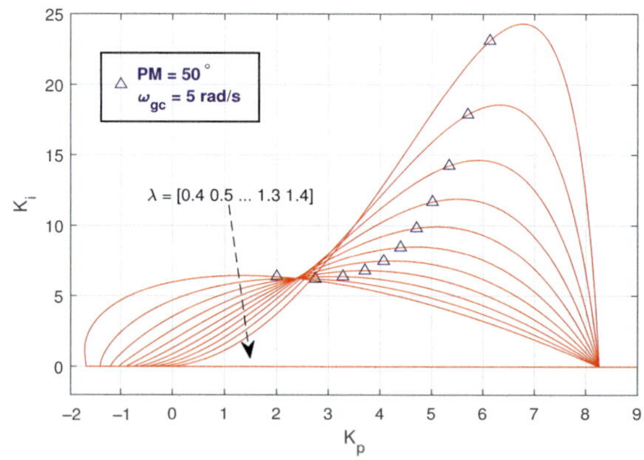

FIGURE 3.4 The designed K_p and K_i satisfying $\omega_{gc} = 5$ rad/s, $\phi_m = 50^\circ$ with different λ and fixed $K_d = 0$.

Step 5. To find the point that satisfies the flat phase constraint, bring the parameters corresponding to all the points in Figure 3.5 into Eq. (3.18). The point that satisfies

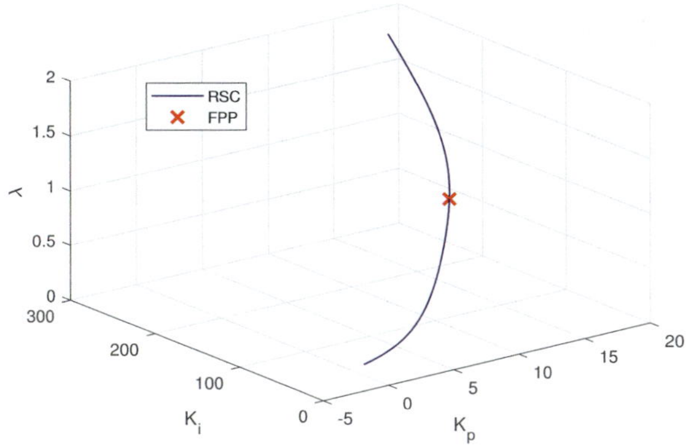

FIGURE 3.5 The RSC and the FPP in three-dimensional parameter space with $K_d = 0$.

the flat phase constraint is shown as a red cross in Figure 3.5, which is the *flat phase point* (FPP). The parameter set (K_p, K_i, λ, K_d) corresponding to the FPP satisfies the ω_{gc}, phase margin and the flat phase constraint.

3.4.1 The Optimal Robust Controller Design

Step 6. Given different K_d, repeat *Step 4* to get different FPP as shown by the red circle in Figure 3.6. Sweeping all the $K_d \in [-1, 1]$, the flat phase constraint can be determined as shown in Figure 3.7. All the points satisfy the prespecified ω_{gc}, ϕ_m and the flat phase constraint.

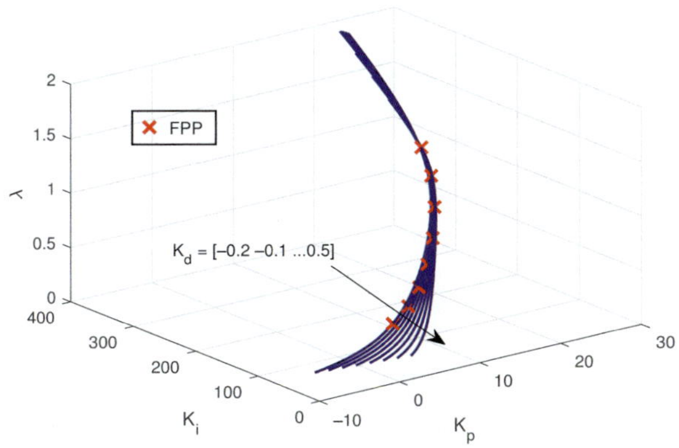

FIGURE 3.6 The RSC and the FPP in the three-dimensional parameter space with different K_d.

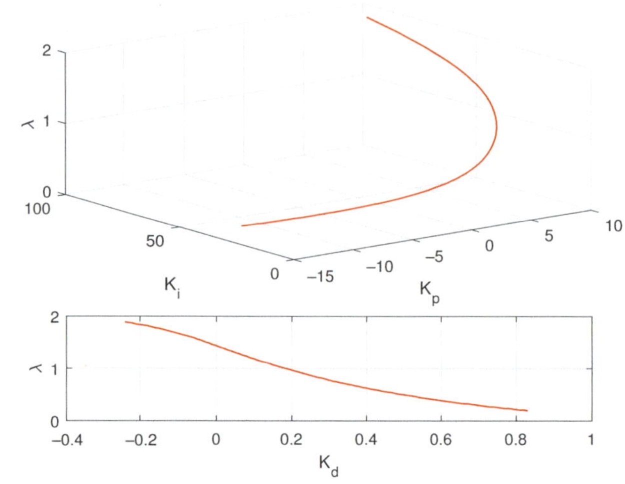

FIGURE 3.7 Flat phase constraint.

Step 7. Selecting every point on the flat phase constraint in turn, J_{ITAE} can be calculated through the step response simulation. The relationship diagram between λ and J_{ITAE} is shown in Figure 3.8, where the smallest J_m point is shown as a red star.

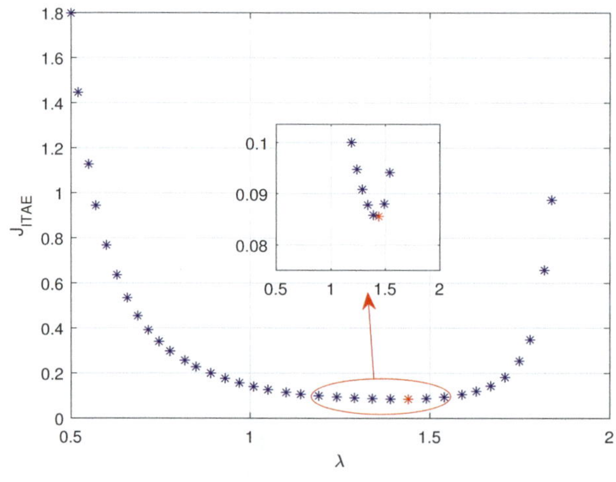

FIGURE 3.8 J_{ITAE} with different FPP on the flat phase constraint.

3.4.2 Complete Achievable Region of ω_{gc} and ϕ_m

In this subsection, the procedures for collecting the complete achievable region of ω_{gc} and ϕ_m are presented.

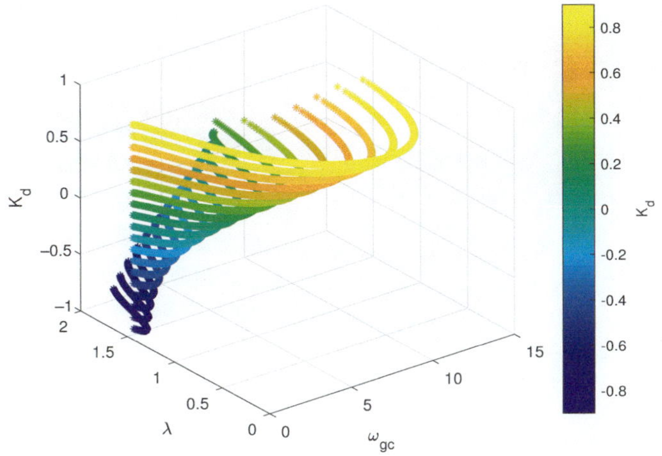

FIGURE 3.9 Achievable region of ω_{gc} with $\phi_m = 50^{\circ}$.

Step 8. According to *Step 3*, the upper boundary ω_{1max} can be determined by sweeping all the λ with fixed K_d. Sweeping over the $\omega \in (0, \omega_{1max}]$, the achievable ω_{gc} satisfying the flat phase constraint and $\phi_m = 50^{\circ}$ with fixed K_d can be determined. Then, sweeping over the $K_d \in [-1, 1]$, the total achievable ω_{gc} can be collected as shown in Figure 3.9. Sweeping all the phase margin $\phi_m \in (0, 2\pi)$, the complete achievable region of ω_{gc} and ϕ_m can be collected as shown in Figure 3.10a with K_d as the vertical axis in the 3D plot. In order to better understand Figure 3.10a, the achievable regions are shown in Figure 3.10b–3.10d with $K_d = -0.4, 0, 0.4$, respectively.

The normalized form of the FOPTD model (3.2) is [33],

$$P(s) = \frac{1}{Ts+1} e^{-(L/T)Ts} = \frac{1}{\bar{s}+1} e^{-\bar{L}\bar{s}}, \tag{3.20}$$

where $\bar{L} = L/T$ and $\bar{L} = Ts$. K in (3.2) can be included in gain of the controller.

Thus, the entire region of ϕ_m and ω_{gc} is obtained as Λ for the normalized form below:

$$P_0(s) = \frac{1}{s+1} e^{-Ls}. \tag{3.21}$$

Thus, the entire achievable region of ϕ_m and ω_{gc} can be conveniently obtained with the T times ω_{gc} axis change of the region Λ. With the procedures in Sec. 3.4, the complete achievable regions can be collected with $L = 1$ and 10 in Figures 3.11a and 3.12a.

3.4.3 Discussion

On the one hand, while $K_d = 0$, $C(s)$ becomes a FOPI controller. Given the prespecified ω_{gc} and ϕ_m, a certain robust FOPI controller can be obtained without the

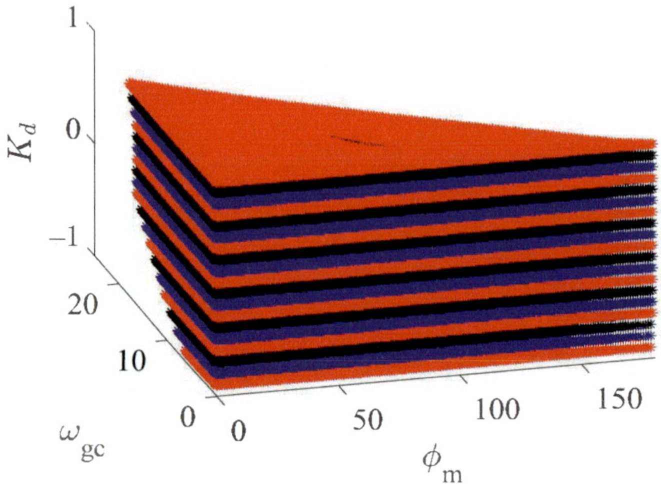

(a) Complete achievable region.

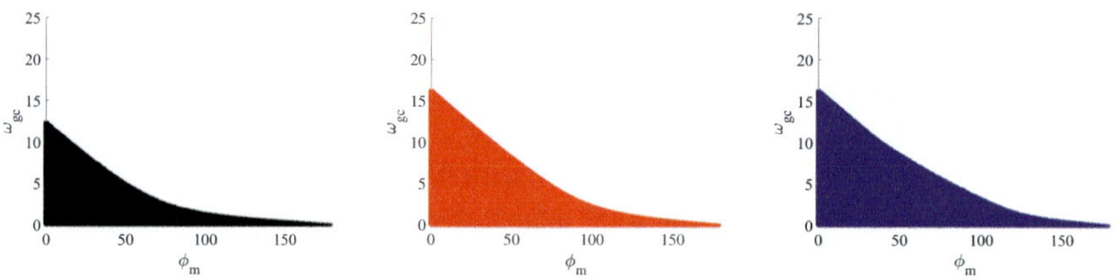

(b) $K_d = -0.4$. **(c)** $K_d = 0$. **(d)** $K_d = 0.4$.

FIGURE 3.10 The achievable region of ω_{gc} and ϕ_m with $L = 0.1$.

time-domain specification. Moreover, the achievable regions can be visualized in the plane as shown in Figures 3.10c, 3.11c, and 3.12c. Comparing with the FOPI controller, the proposed FOPI$^\lambda$D controller, i.e. $K_d \neq 0$, can achieve the optimal control performance because in addition to meeting the frequency-domain specifications, it also needs to minimize the time-domain specifications (ITAE). Furthermore, the complete achievable region of ω_{gc} and ϕ_m for the proposed FOPI$^\lambda$D is significantly bigger than that for the FOPI controller according to the Figures 3.10a, 3.11a, and 3.12a.

On the other hand, in order to show the advantages of FOPI$^\lambda$D over IOPID more clearly, the complete achievable region with fractional order λ as the vertical axis in 3D plot with $L = 0.1, 1, 10$ are shown in Figures 3.13a, 3.14a, and 3.15a, respectively. For the IOPID controller, it has been found that the control becomes more and more challenging as L increases, which is corresponding to the achievable regions decreasing with λ increasing in Figures 3.13c, 3.14c, and 3.15c. However, the optional combinations of ω_{gc} and ϕ_m for the proposed FOPI$^\lambda$D controller are significantly improved over those of the IOPID controller.

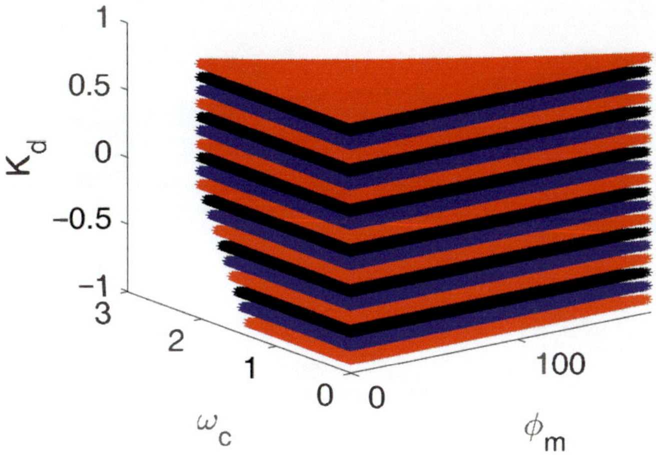

(a) Complete achievable region.

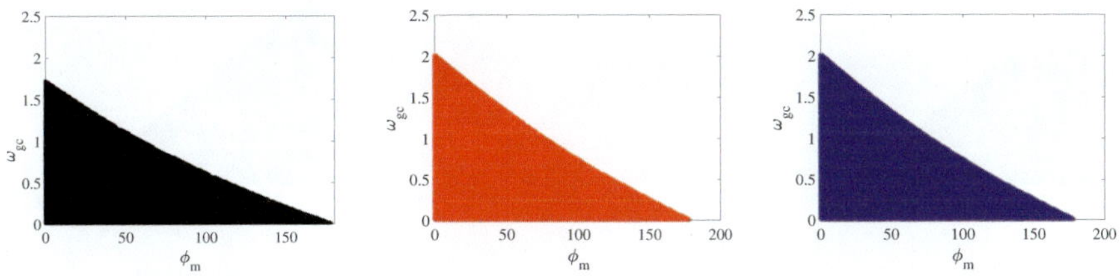

(b) $K_d = -0.4$.   **(c)** $K_d = 0$.   **(d)** $K_d = 0.4$.

FIGURE 3.11 The achievable region of ω_{gc} and ϕ_m with $L = 1$.

3.5 SIMULATION RESULTS

To verify the advantages of the proposed FOPI$^\lambda$D and design procedures, the control performances using the FOPI$^\lambda$D controllers are compared with those of the FOPI [33], TPFOPID [24], ZN-FOPID [51], FF-FOPID [47] and SIMC-PI controllers [52] by simulation.

3.5.1 Flat Phase Verification

In this subsection, the robustness to loop gain variations is verified by comparison with the designed control system without flat phase constraint.

Given the FOPTD system (3.22) and the frequency-domain specifications $\omega_{gc} = 1$ rad/s and $\phi_m = 60°$, the FOPI$^\lambda$D controller can be designed as (3.23) with the proposed method,

$$P(s) = \frac{1}{s+1} e^{-0.1s}, \tag{3.22}$$

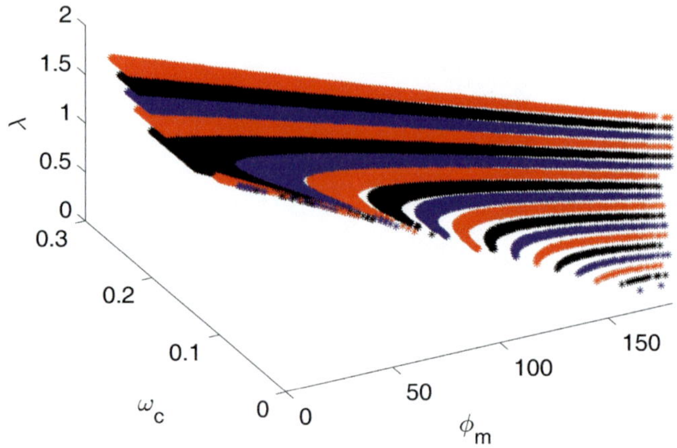

(a) Complete achievable region.

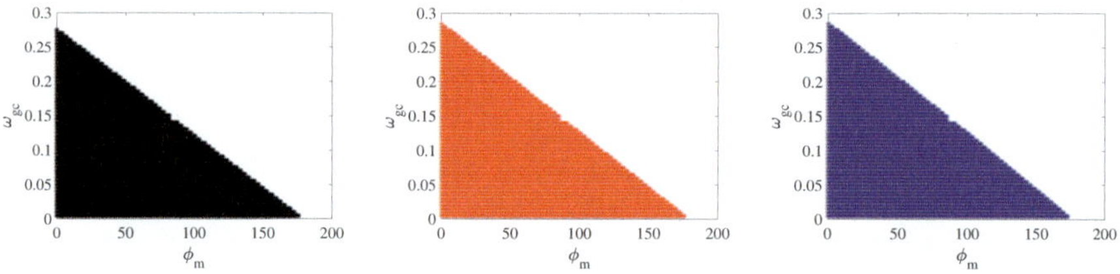

(b) $K_d = -0.4$. **(c)** $K_d = 0$. **(d)** $K_d = 0.4$.

FIGURE 3.12 The achievable region of ω_{gc} and ϕ_m with $L = 10$.

$$C_{FOPI^{\lambda}D}(s) = 0.80371 + \frac{1.2985}{s^{1.15}} - 0.06s. \tag{3.23}$$

One FOPI$^{\lambda}$D controller without flat phase constraint can be designed to compare with that of the flat phase constraint. For example, fixed $\lambda = 0.9$ and $K_d = 0.2$, two controller parameters K_p and K_i can be designed according to two frequency-domain specifications ω_{gc} and ϕ_m, and the designed controller is

$$C_{FOPI^{\lambda}D}(s) = 0.25941 + \frac{1.5416}{s^{0.9}} + 0.2s. \tag{3.24}$$

The open-loop Bode diagrams of two systems are presented in Figure 3.16. Clearly, both control systems fulfill the frequency domain specification ω_{gc} and ϕ_m, one control system satisfies the flat phase constraint, and another one has a slope at the gain crossover frequency.

To show the robustness to the process gain variations, the step responses are presented of two designed control systems with the system loop gains as K = 0.6, 0.8,

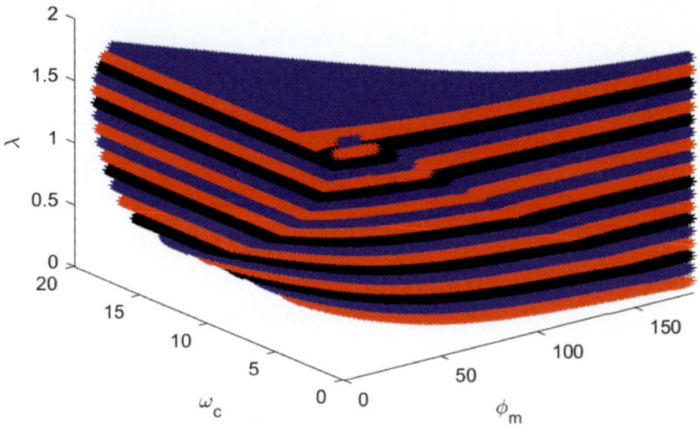

(a) Complete achievable region.

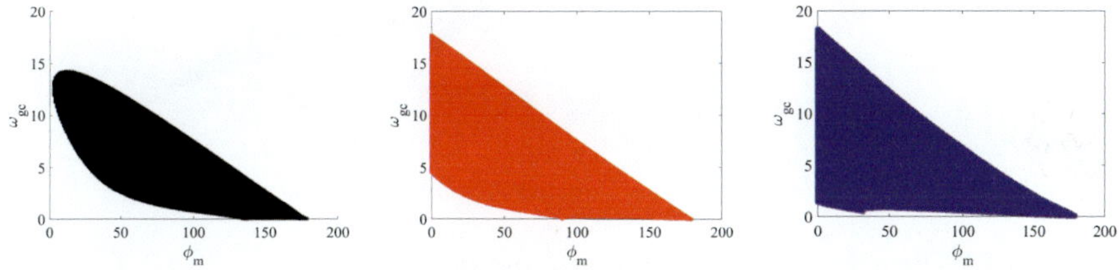

(b) $\lambda = 0.5$. **(c)** $\lambda = 1$. **(d)** $\lambda = 1.5$.

FIGURE 3.13 The achievable region of ω_{gc} and ϕ_m with $L = 0.1$.

1.0, 1.2, and 1.4, and the results are shown in Figure 3.17a and 3.17b of two control systems.

From Figures 3.17a and 3.17b, it can be observed that in the control system without the flat phase phase constraint, the overshoot varies significantly with changes in the system gain. The designed control system with the flat phase constraint keeps the overshoots constant with the same variations of the plant gain.

3.5.2 Illustration with Examples

3.5.2.1 Example 1

Given the FOPTD system (3.25) and the frequency-domain specifications $\omega_{gc} = 5$ rad/s and $\phi_m = 58°$, the proposed FOPI$^\lambda$D is compared with FOPI$^\lambda$ [33] and TPFOPID [24] controllers.

The FOPI$^\lambda$D controller obtained with our proposed method is

$$C_{FOPI^\lambda D1}(s) = 6.519 + \frac{25.0716}{s^{1.47}} + 0.09s. \tag{3.25}$$

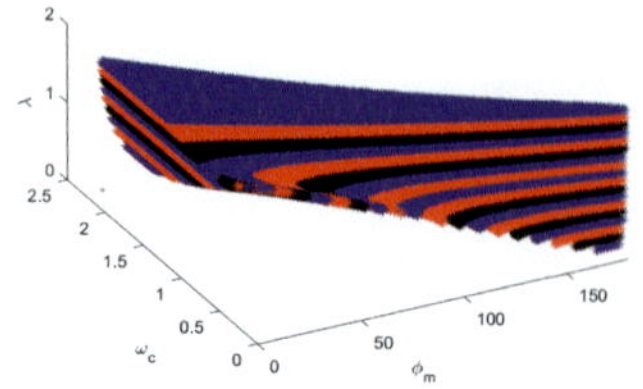

(a) Complete achievable region.

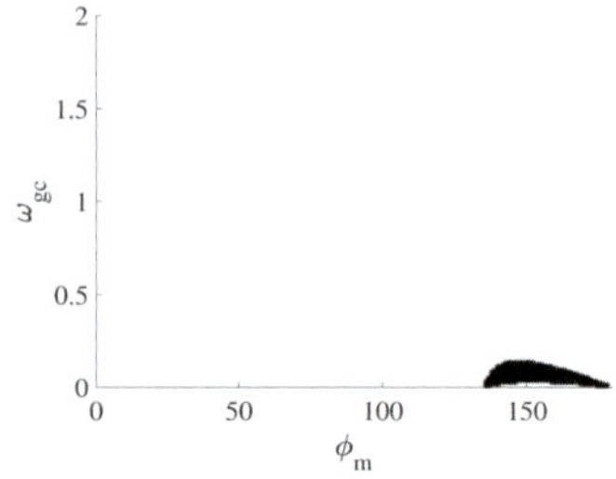

(b) Achievable region of $\lambda = 0.5$.

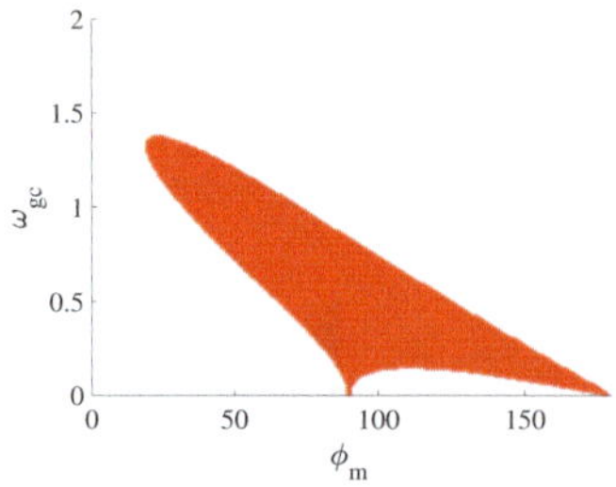

(c) Achievable region of $\lambda = 1$.

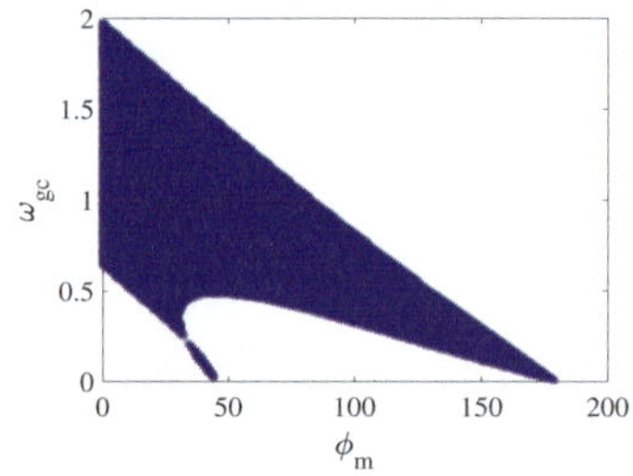

(d) Achievable region of $\lambda = 1.5$.

FIGURE 3.14 The achievable region of ω_{gc} and ϕ_m with $L = 1$.

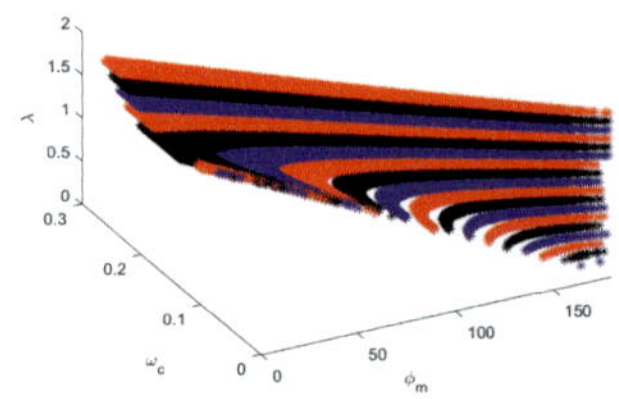

(a) Complete achievable region.

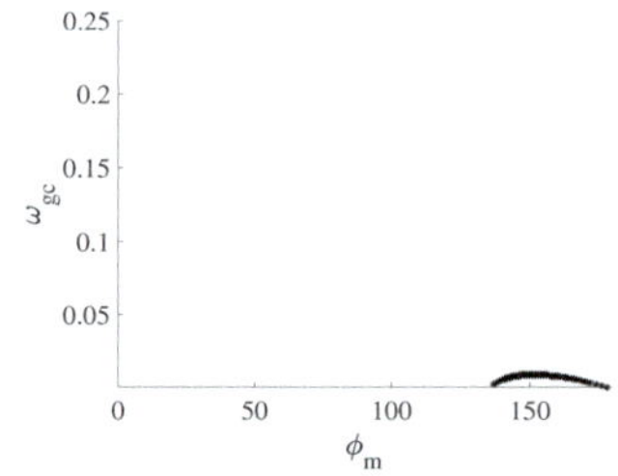

(b) Achievable region of $\lambda = 0.5$.

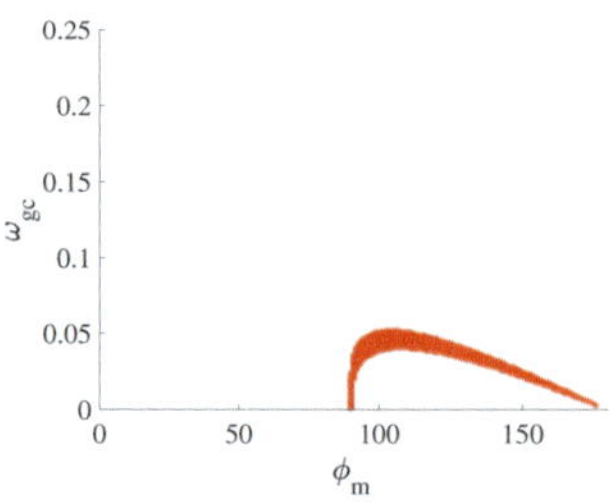

(c) Achievable region of $\lambda = 1$.

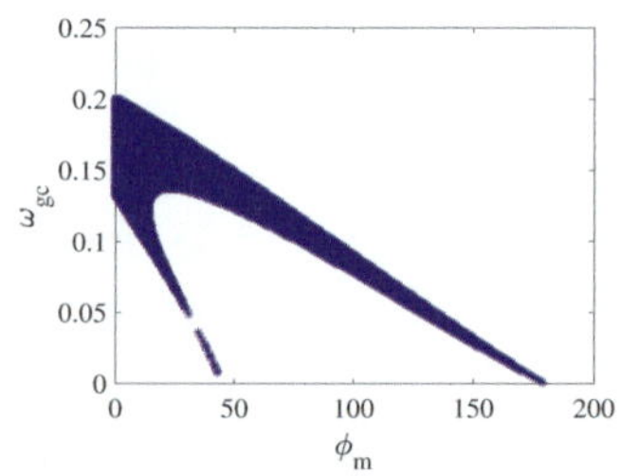

(d) Achievable region of $\lambda = 1.5$.

FIGURE 3.15 The achievable region of ω_{gc} and ϕ_m with $L = 10$.

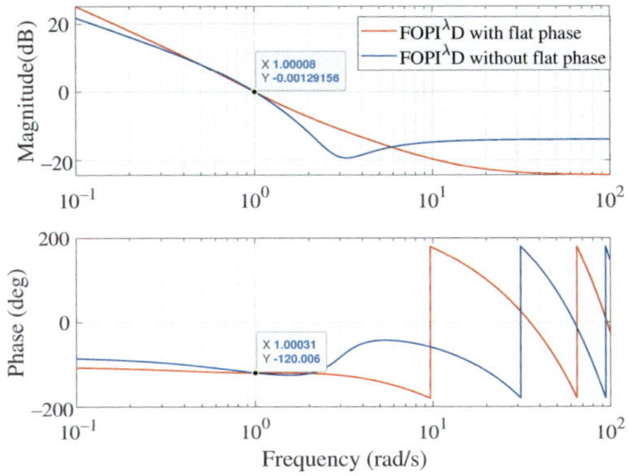

FIGURE 3.16 Open-loop Bode plots of FOPI$^{\lambda}$D control systems with and without flat phase constraint.

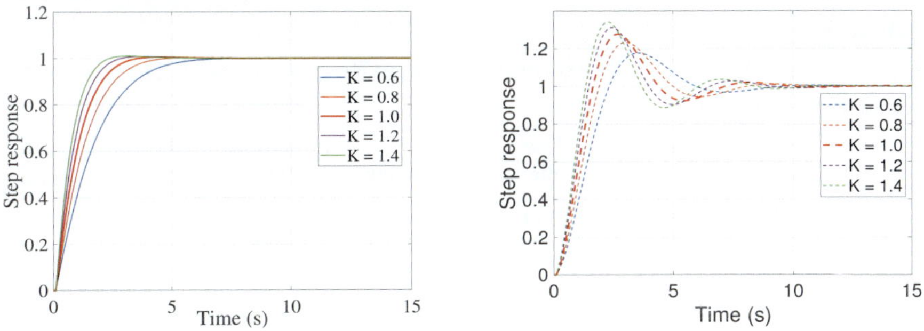

(a) With flat phase constraint.

(b) Without flat phase constraint.

FIGURE 3.17 The step responses of FOPI$^{\lambda}$D control system with and without flat phase constraint.

Under the same frequency-domain specification $\omega_{gc} = 5$ rad/s and $\phi_m = 58°$, the FOPI and TPFOPID controllers are designed, respectively, as

$$C_{FOPI1}(s) = 8.0489 + \frac{56.385}{s^{1.75}}, \tag{3.26}$$

$$C_{TPFOPID1}(s) = 4.0968(1 + \frac{3.4057}{s^{0.8755}} + 0.08909 s^{0.8755}). \tag{3.27}$$

The open-loop Bode plots of three systems are presented in Figure 3.18a. The step and disturbance rejection responses of closed-loop systems using the FOPI$^{\lambda}$D controller $C_{FOPI^{\lambda}D1}(s)$, the FOPI controller $C_{FOPI1}(s)$, and the TPFOPID controller $C_{TPFOPID1}(s)$ are presented in Figure 3.18b. The comparison results about frequency- and time-domain characteristics are summarized in Table 3.1.

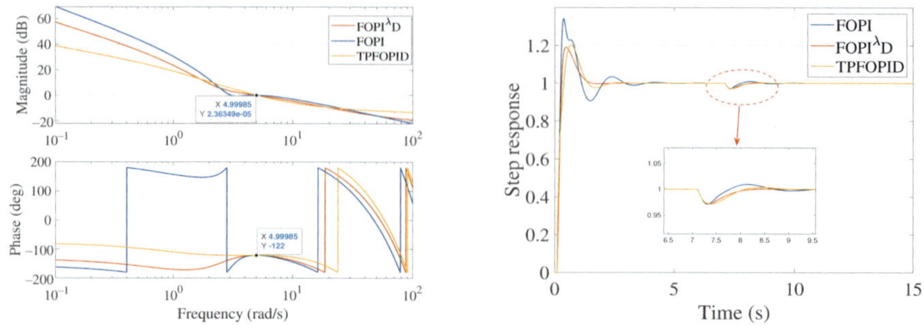

(a) Open-loop Bode plots. **(b)** The step and load disturbance responses.

FIGURE 3.18 The comparison of open-loop Bode plot and step responses using the designed FOPI, TPFOPID, and FOPI$^\lambda$D controllers (Example 1).

TABLE 3.1
Comparison of Step and Disturbance Rejection Responses with Three Control Systems (Example 1)

	ω_{gc} **(rad/s)**	$\phi_m(°)$	**Rise Time(s)**	**Overshoot(%)**	**Settling Time(s)**	**Speed Drop(%)**	J_{ITAE}
FOPI	5	58	0.2305	35.4	2.682	2.87	0.3114
TPFOPID	5	58	0.373	20.3	1.877	2.83	0.1401
FOPI$^\lambda$D	5	58	0.2785	19	1.237	2.87	0.09357

Figure 3.18b and Table 3.1 show that the frequency-domain specifications of three systems are the same and all fulfill the given frequency-domain specifications. And the designed FOPI$^\lambda$D controller achieves the smallest overshoot (19%) and the shortest settling time (1.237 s) comparing with FOPI and TPFOPID controllers. Obviously, the designed FOPI$^\lambda$D system achieves the best performance from the values of J_{ITAE}. Meanwhile, all three control systems achieve the same performance in terms of anti-load disturbance with similar speed drop.

3.5.2.2 Example 2

Considering the same process (3.25) with different frequency-domain design specifications $\omega_{gc} = 2$ rad/s and $\phi_m = 55°$, the FOPI$^\lambda$D controller can be designed as (3.28) with the proposed method,

$$C_{FOPI^\lambda D2}(s) = 1.929 + \frac{3.4057}{s^{1.22}} - 0.17s. \tag{3.28}$$

With the same frequency-domain specifications ω_{gc} and ϕ_m, the FOPI and TPFOPID controllers obtained using the proposed method in [33] and [24] are given, respectively, as

$$C_{FOPI2}(s) = 1.6236 + \frac{3.6236}{s^{1.07}}, \tag{3.29}$$

$$C_{TPFOPID2}(s) = 1.364(1 + \frac{2.826}{s^{0.9795}} + 0.08846s^{0.9795}). \tag{3.30}$$

The Bode diagrams of three systems are presented in Figure 3.19a. The step and disturbance rejection responses of closed-loop systems using the FOPI$^{\lambda}$D controller $C_{FOPI^{\lambda}D2}(s)$, the FOPI controller $C_{FOPI2}(s)$, and the TPFOPID controller $C_{TPFOPID2}(s)$ are presented in Figure 3.19b. The comparison results about frequency-domain and time-domain characteristics are summarized in Table 3.2.

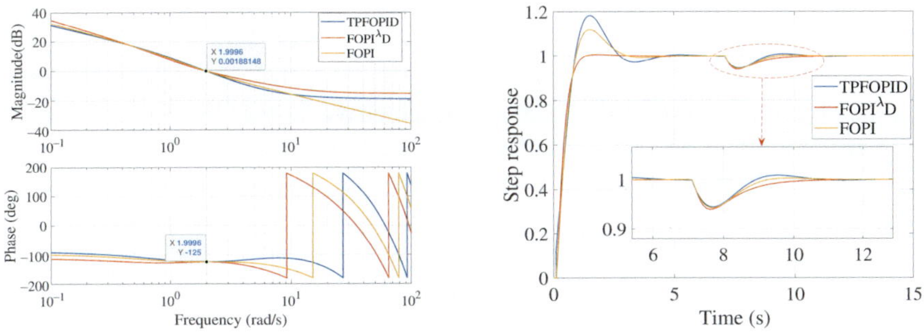

(a) Open-loop Bode plots.

(b) The step and load disturbance responses.

FIGURE 3.19 The comparison of open-loop Bode plot and step responses using the designed FOPI, TPFOPID, and FOPI$^{\lambda}$D controllers (Example 2).

TABLE 3.2
Comparison of Step Responses with Three Control Systems (Example 2)

	ω_{gc} (rad/s)	ϕ_m (°)	Rise Time(s)	Overshoot(%)	Settling Time(s)	Speed Drop(%)	J_{ITAE}
FOPI	2	55	0.943	12	2.695	5.65	0.3957
TPFOPID	2	55	0.919	18.1	3.849	5.5	0.6152
FOPI$^{\lambda}$D	2	55	1.388	0.7	1.388	5.98	0.1396

It can be easily seen from Figure 3.19b and Table 3.2 that the frequency-domain specifications of three systems are the same and all fulfill the given frequency-domain specifications $\omega_{gc} = 2$ rad/s and $\phi_m = 55°$. And the designed FOPI$^{\lambda}$D controller achieves the smallest overshoot (0.7 %) and the shortest settling time (1.388 s) compared with FOPI and TPFOPID controllers, though it has the longest rise time

(1.388 s). According to the values of J_{ITAE} of the three control systems, the designed FOPI$^\lambda$D system achieves optimal performance. In terms of the disturbance rejection, the performances of the three control systems are slightly different.

3.5.2.3 Example 3

In [51], Duarte Valério proposed an analytical method for tuning FOPID controller with Ziegler–Nichols-type rules (ZN-FOPID). In this subsection, the designed FOPI$^\lambda$D controller is compared with the ZN-FOPID controller. Given the FOPTD system (3.31),

$$P(s) = \frac{1}{s+1}e^{-0.2s}. \tag{3.31}$$

The ZN-FOPID controller can be designed as

$$C_{ZN-FOPID}(s) = 2.2719 + \frac{1.2168}{s^{1.1441}} - 0.7589s^{0.0107}. \tag{3.32}$$

The Bode plot of the ZN-FOPID system is presented as the blue line in Figure 3.20a. The frequency-domain specifications are $\omega_{gc} = 1.248$ rad/s and $\phi_m = 78.7°$. For fair comparison, the FOPI$^\lambda$D is designed with the same ω_{gc} and ϕ_m. The FOPI$^\lambda$D can be designed as follows,

$$C_{FOPI^\lambda D3}(s) = 1.4034 + \frac{1.407}{s^{1.06}} + 0.14s. \tag{3.33}$$

The responses of frequency-domain and time-domain are presented in Figure 3.20a and 3.20b. The comparison results about frequency-domain and time-domain characteristics are summarized in Table 3.3.

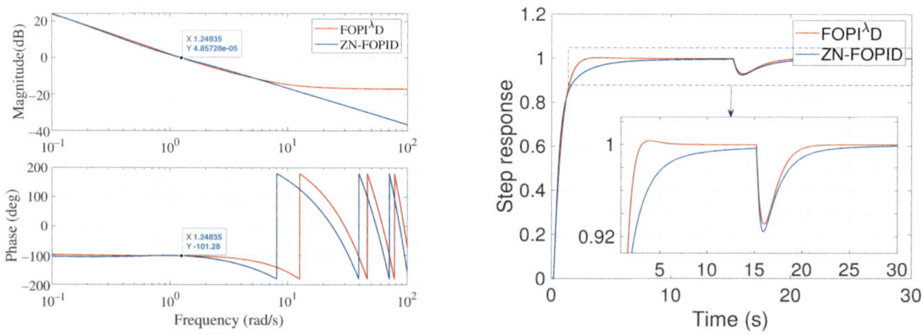

(a) Open-loop Bode plots.

(b) The step and load disturbance responses.

FIGURE 3.20 The comparison of open-loop Bode plot and step responses using the designed ZNFOPID and FOPI$^\lambda$D controllers (Example 3).

TABLE 3.3
Comparison of Step Responses with Two Control Systems (Example 3)

	ω_{gc} (rad/s)	$\phi_m(^\circ)$	Rise Time(s)	Overshoot(%)	Settling Time(s)	Speed Drop(%)	J_{ITAE}
FOPI$^{\lambda}$D	1.248	78.7	2.325	0.3	2.325	6.88	5.134
ZN-FOPID	1.248	78.7	5.357	0	5.357	7.54	22.63

According to Figure 3.20b and Table 3.3, the frequency-domain specifications of two systems are the same. Two systems both have small overshoot, but the settling time of the designed FOPI$^{\lambda}$D system is less than designed ZN-FOPID system. Clearly, the designed FOPI$^{\lambda}$D system achieves better control performance according to the values of the J_{ITAE}. In terms of anti-load disturbance, the designed ZN-FOPID control system has larger speed drop and spends longer time to reach tracking value.

3.5.2.4 Example 4

Yumuk et al. [47] proposed a FF-FOPID controller based on Bode's ideal. In this subsection, the proposed FOPI$^{\lambda}$D controller is compared with FF-FOPID controller. Given the FOPTD system (3.34) and the frequency-domain design specifications $\omega_{gc} = 0.5$ rad/s and $\phi_m = 60^\circ$,

$$P(s) = \frac{1}{0.8s+1}e^{-s}. \tag{3.34}$$

The FOPI$^{\lambda}$D controller can be designed as (3.35) using the proposed the method

$$C_{FOPI^{\lambda}D4}(s) = 0.71451 + \frac{0.535}{s^{1.18}} + 0.31s. \tag{3.35}$$

and the FF-FOPID controller is obtained as (3.36) using the method in [51],

$$C_{FF-FOPID}(s) = 0.3959s^{-0.01502}(1 + \frac{1.25}{s}). \tag{3.36}$$

The open-loop Bode diagrams of two systems are presented in Figure 3.21a. The step and disturbance rejection responses using the FOPI$^{\lambda}$D $C_{FOPI^{\lambda}D4}(s)$ and the FF-FOPID controller $C_{FF-FOPID}(s)$ are presented in Figure 3.21b. The comparison results about frequency-domain and time-domain characteristics are summarized in Table 3.4.

According to Figure 3.21b and Table 3.4, the frequency-domain specifications of the two systems are the same. In terms of time-domain characteristics, the designed FOPI$^{\lambda}$D system demonstrates better performance than the FF-FOPID system: the FOPI$^{\lambda}$D system has no overshoot, but the overshoot of the FF-FOPID system is

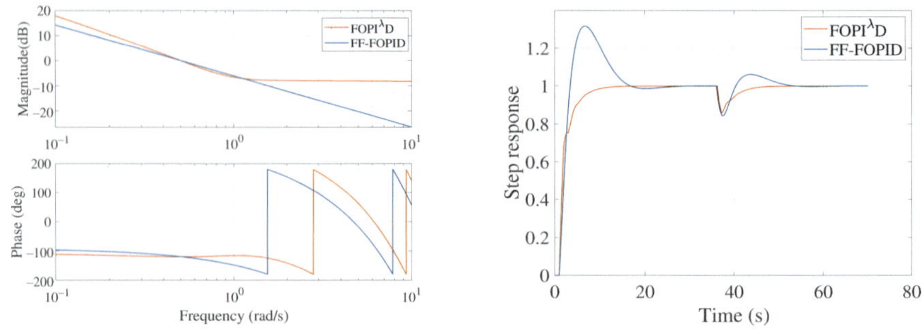

(a) Open-loop Bode plots. **(b)** The step and load disturbance responses.

FIGURE 3.21 The comparison of open-loop Bode plot and step responses using the designed FFFOPID and FOPI$^\lambda$D controllers (Example 4).

TABLE 3.4

Comparison of Step Responses with Two Control Systems (Example 4)

	ω_{gc} (rad/s)	$\phi_m(^\circ)$	Rise Time(s)	Overshoot(%)	Settling Time(s)	Speed Drop(%)	J_{ITAE}
FOPI$^\lambda$D	0.5	60	9.034	0	9.034	16.27	0.496
FF-FOPID	0.5	60	3.42	31.7	15.34	16.71	2.044

31.7 %, and the FOPI$^\lambda$D system achieves less settling time (9.034 s) than the FF-FOPID system. According to the values of J_{ITAE}, the FOPI$^\lambda$D system achieves better control performance than the FF-FOPID system. Moreover, the FF-FOPID system has a larger speed drop (16.71 %) and spends a longer time to reach tracking value.

3.5.2.5 Example 5

Given the process (example E3 in [52])

$$P(s) = \frac{2(15s+1)}{(20s+1)(s+1)(0.1s+1)^2}. \tag{3.37}$$

The plant (3.38) can be approximated as a FOPTD process using the half rule [52],

$$P_1(s) = \frac{1.5}{1.05s+1}e^{-0.15s}. \tag{3.38}$$

According to the Skogestad Internal Model Control (SIMC) design method, the SIMC-PI [52] controller can be designed as,

$$C_{SIMC-PI}(s) = 2.33(1+\frac{1}{1.05s}). \tag{3.39}$$

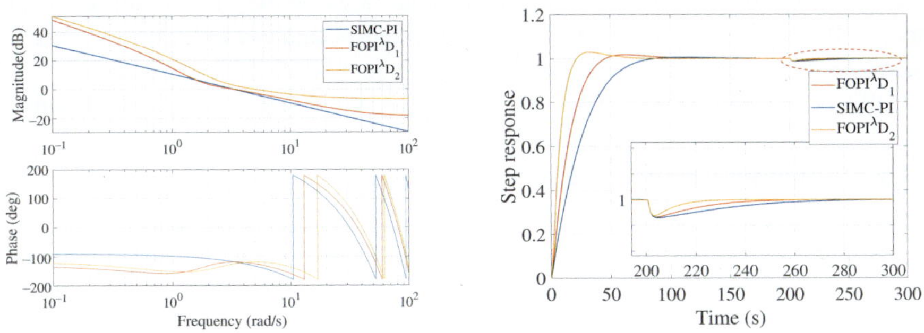

(a) Open-loop Bode plots. **(b)** The step responses.

FIGURE 3.22 The comparison of open-loop Bode plot and step responses using the designed SIMC-PI and FOPI$^\lambda$D controllers (Example 5).

The Bode diagram of the SIMC-PI system is presented as a blue line in Figure 3.22a. The frequency-domain specifications are $\omega_{gc} = 3.329$ rad/s and $\phi_m = 61.4°$. To show flexibility of the proposed FOPI$^\lambda$D controller design strategy, two FOPI$^\lambda$D controllers were designed with different frequency-domain specifications: (1) the same ω_{gc} and ϕ_m; and (2) higher ω_{gc} (1.5 times) and the same ϕ_m. Two controllers can be designed as

$$C_{FOPI^\lambda D_1}(s) = 3.077 + \frac{6.672}{s^{1.43}} + 0.08s. \tag{3.40}$$

$$C_{FOPI^\lambda D_1}(s) = 4.228 + \frac{12.229}{s^{1.29}} + 0.32s. \tag{3.41}$$

According to Figure 3.22b and Table 3.5, the rise time and settling time of the two designed FOPI$^\lambda$D control systems are both much less than that of the SIMC-PI one. The overshoot of the SIMC-PI control is smaller than other two, but all of three response overshoots are fully acceptable. It can be clearly seen that the two designed FOPI$^\lambda$D systems achieve much better control performance than SIMC-PI control system according to the values of the J_{ITAE}. In terms of anti-load disturbance, the designed SIMC-PI control system has the biggest speed drop (1.93 %) and spends

TABLE 3.5
Comparison of Step Responses with Three Control Systems (Example 5)

	ω_{gc} (rad/s)	$\phi_m(°)$	Rise Time(s)	Overshoot(%)	Settling Time(s)	Speed Drop(%)	J_{ITAE}
SIMC-PI	3.329	61.4	86.86	150.1	0.5	1.39	462.7
FOPI$^\lambda$D$_1$	3.329	61.4	45.83	112.5	1.8	1.34	203.5
FOPI$^\lambda$D$_2$	4.994	61.4	21.93	76.12	2.9	1.23	72.35

the longest time to recover to the tracking value. Meanwhile, the flexibility of the proposed FOPI$^\lambda$D control design is also illustrated in this example; using a higher gain cross-over frequency design specification, higher bandwidth can be achieve for the FOPI$^\lambda$D control system to get better tracking and anti disturbance performance.

3.5.2.6 Implementation for Real System

The advantages of the proposed FOPI$^\lambda$D controller have been verified based on the above simulation illustrations. Note that ideal derivative is adopted for simulation. However, the ideal derivative can amplify the measurement noise which exists in real systems. In general, the filtered derivative can be used to replace the ideal derivative for the practical systems, i.e. s^α is replaced by $\frac{s^\alpha}{K_d/Ns^\alpha+1}$, where N is the filter coefficient [55].

To explain clearly, the designed control systems ($C_{FOPI^\lambda D1}(s)$) in Example 1 is taken as an example. The Bode plot comparison between the designed $C_{FOPI^\lambda D1}(s)$ system with ideal derivative and that with filtered derivative ($N = 5$) is presented in Figure 3.23a. Clearly, the prespecific ω_{gc} and ϕ_m is basically the same with only minor variations (gain margin from 58° to 58.1° and gain crossover frequency from 5 rad/s to 5.059 rad/s). Further, the comparison of step responses between the designed $C_{FOPI^\lambda D1}(s)$ system with filtered derivative and the designed $C_{FOPI1}(s)$ system is presented in Figure 3.23b. According to Figure 3.23b, even though FOPI$^\lambda$D is analyzed with the ideal derivative, the performance is also satisfactory with the filtered derivative and the same parameters. Therefore, the proposed FOPI$^\lambda$D designed method can be fully recommended with measurement noise in real systems.

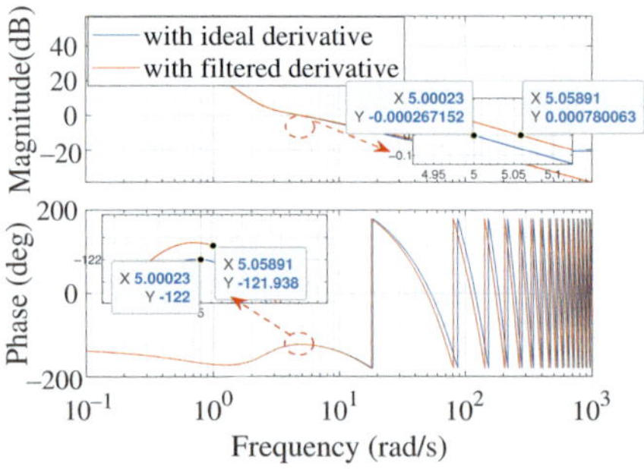

(a) Open-loop Bode plots.

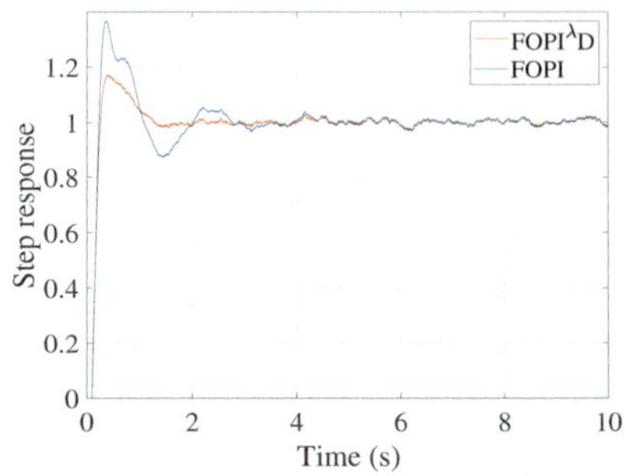

(b) The step responses.

FIGURE 3.23 The comparison of open-loop Bode plot with ideal derivative and filtered derivative and step responses using the designed FOPI and FOPI$^\lambda$D controller.

3.6 CHAPTER SUMMARY

This chapter proposes a design method for FOPI$^\lambda$D to achieve both frequency-domain specifications (i.e. ω_{gc}, ϕ_m and flat phase constraint) and time-domain requirements (i.e. ITAE) for all the FOPTD systems. Moreover, the complete feasible regions of two frequency-domain specifications can be collected for the robust FOPI$^{\lambda}$D controller design with a synthesis scheme. This region illustrates all the combinations of ω_{gc} and ϕ_m of a robust FOPI$^{\lambda}$D controller. The advantage of FOPI$^{\lambda}$D controller is clearly shown from the comparisons of the feasible regions with the FOPI controller and IOPID controller. Simulation illustration is performed for delay-dominant systems, lag-dominant systems, and high-order process with one zero to show the advantages on control performance of the designed optimal robust FOPI$^{\lambda}$D controller, comparing with the FOPI controller, TPFOPID controller, ZN-FOPID controller, FF-FOPID controller, and SIMC-PI controller.

4 Five-Parameter FO-PI$^{\lambda}$D$^{\mu}$ Control

4.1 INTRODUCTION

In Chapters 2 and 3, we have discussed the design methodologies for three-parameter fractional order PID and four-parameter fractional order PID controllers. Furthermore, methods for collecting reachable regions of the frequency-domain index have been provided. In fact, Podlubny [9] introduced a concept of a fractional order PI$^{\lambda}$D$^{\mu}$ (FOPI$^{\lambda}$D$^{\mu}$) controller, involving fractional order integrator and fractional order differentiator, which has five unknown parameters.

The proposed FOPI$^{\lambda}$D$^{\mu}$ was demonstrated with the better control performance, compared to the typical PID controller. However, the tuning method based on the frequency-domain specifications becomes more complex because of the extra fractional orders λ and μ of the integrator and differentiator. Thus, aiming to the proposed FOPI$^{\lambda}$D$^{\mu}$ controller by Podlubny, two research questions arise:

- ▲ Question 1: How do we determine the parameters of FOPI$^{\lambda}$D$^{\mu}$ controller through the frequency-domain analysis? To the best of our knowledge, there is no comprehensive analytical method available to design a FOPI$^{\lambda}$D$^{\mu}$ controller with five parameters to satisfy user-specified five frequency-domain specifications. Therefore, it is meaningful to design FOPI$^{\lambda}$D$^{\mu}$ controllers theoretically where all their five parameters can be determined uniquely under five frequency-domain constraints: (i) specified phase margin; (ii) specified gain crossover frequency; (iii) specified gain margin; (iv) specified phase crossover frequency; and (v) flat phase constraint.
- ▲ Question 2: How do we know a picked combination of phase margin, gain crossover frequency, gain margin, and phase crossover frequency is achievable for a stabilizing controller while maintaining the flat phase feature? It is desirable to obtain a four-dimensional graph for checking the complete set of the achievable phase margin, gain crossover, gain margin, and phase crossover before the controller design.

The PID controller is still the most widely applied in industrial process control field, despite the rapid development of the advanced controls [56, 57]. Over the years, PID controller tuning methods based on frequency-domain specifications have been a hot topic, and many scholars have made significant contributions [34, 58–66]. In [58], Weng et al. proposed the PID/proportional integral (PI) controller design formula for first-order and second-order plus time-delay systems based on ϕ_m and A_m. Fung et

DOI: 10.1201/9781003397632-6

al. [60] proposed a PI controller tuning method satisfying user-specified ϕ_m and A_m by transforming the issue of solving a set of nonlinear equations into determining the intersection of two curves. All the above studies are based on gain margin and phase margin, which reflect the relative stability of the control system. The rise time of the step response is another performance indicator of the control system, which is related to gain crossover frequency or bandwidth. In [34], Chen et al. proposed a PID controller design method, meeting the given ω_{gc}, ϕ_m and an additional condition that the phase Bode plot is flat at the ω_{gc}, which ensures that the control system is robust to gain variations and the overshoot of the response is almost constant within a gain variation range. Mikhalevich et al. [64] proposed a PID tuning method based on desired phase margin. Najafizadegan et al. [65] proposed an internal model control–PID controller by open-loop shaping to satisfy the desired gain crossover frequency and phase margin.

Zhao et al. [67] proposed a FOPID controller design method for fractional order models and demonstrated that the FOPID controller achieves better control performance compared to the classical PID controller. The designed control system satisfies user-desired A_m, ϕ_m, and their corresponding crossover frequencies, but the fractional orders are determined by trial and error. Monje et al. [68] proposed a method for tuning a PI$^{\lambda}$D$^{\mu}$ controller to fulfill five frequency-domain specifications (ϕ_m, ω_{gc}, robustness to gain variations, high-frequency noise rejection, and low-frequency output disturbance rejection). The designed control system approximately meets the given frequency-domain index and the success of the design method depends on the initial values of the controller parameters. Some researchers focus on the fractional order controllers with three parameters, e.g., the fractional order PI (FOPI) controller, fractional order PD (FOPD) controller, fractional order [PD] (FO[PD]) controller, etc., and put forward a tuning method based on phase margin, gain crossover frequency, and robustness to gain variations for fractional order plants and integer order plants [12, 15, 16, 24]. However, there is no comprehensive analytical method available to design a FOPID controller with five parameters to satisfy user-specified five frequency-domain specifications.

In this chapter, an analytical synthesis for the FOPI$^{\lambda}$D$^{\mu}$ controller with five parameters tuning based on five frequency-domain specifications is proposed for the first order plus normalized time delay (FOPNTD) systems. With the user-specified phase and gain margins, the corresponding the crossover frequencies, and flat phase constraint, a unique FOPI$^{\lambda}$D$^{\mu}$ controller can be determined using the proposed design method. Phase margin and gain margin guarantee the stability and also damping of the control system. Gain crossover frequency meets the desired control bandwidth, which represents the tracking response speed of the control system. The flat phase constraint keeps the system response with the iso-damping property. Moreover, the influence of phase crossover frequency on the loop-shaping of higher-frequency, lower-frequency bands and the resonance peak is presented through frequency-domain analysis with simulation illustration. Furthermore, the achievable regions of four specifications are collected, which is important a priori information before the FOPI$^\lambda$D$^\mu$ controller design. Simulation results demonstrate that the proposed analytical design synthesis for the FOPI$^\lambda$D$^\mu$ controller can achieve better control performance compared with the optimal controllers in Chen et al. [34], Chevalier et al. [24], Yumuk et al. [47] and Valrio et al. [51].

4.2 SYNTHESIS OF FOPI$^\lambda$D$^\mu$ ANALYTICAL DESIGN

The controlled plant targets the FOPNTD systems, which reflect most of the industrial process [57]. The transfer function is expressed as (3.22). The considered controller is a FOPI$^\lambda$D$^\mu$ controller, which is a generalization of the classical PID. The FOPI$^\lambda$D$^\mu$ controller can be written as

$$C(s) = k_p + \frac{k_i}{s^\lambda} + k_d s^\mu, \tag{4.1}$$

where k_p, k_i, and k_d are the proportional, integral, and differential gains ,respectively, the real number λ and μ are the fractional orders, and $\lambda, \mu \in (0,2)$.

To design the FOPI$^\lambda$D$^\mu$ controller, the control system is depicted as presented in Figure 4.1, where r is reference input, y is actual output, d is external disturbance, $C(s)$ is FOPI$^\lambda$D$^\mu$ (4.1), $P(s)$ is the FOPNTD (3.21), and $M_T(A,\phi)$ is a gain-phase margin tester (GPMT), which can provide data for drawing the boundaries of given phase margin and gain margin [29]. The transfer function of $M_T(A,\phi)$ is given as (2.3).

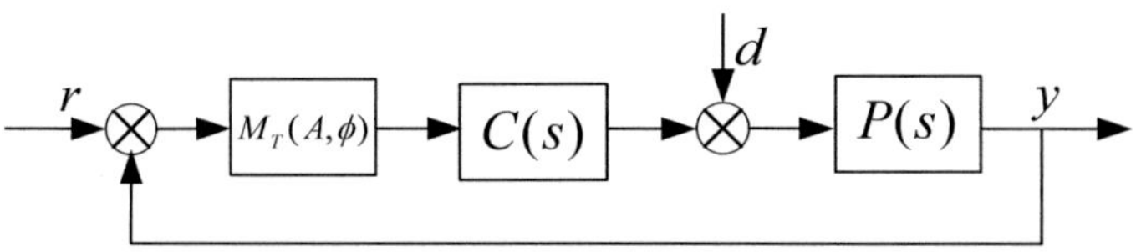

FIGURE 4.1 Feedback control system.

In this chapter, a synthesis scheme based on five frequency-domain specifications for the FOPI$^\lambda$D$^\mu$ controller analytical design is proposed. The five specifications are as follows:

i. Gain crossover frequency (ω_{gc}), which is expressed as (3.11).

ii. Phase margin (ϕ_m), which is expressed as (3.12).

iii. Phase crossover frequency (ω_{pc}), which affects the loop-shaping in high- and low-frequency bands as to be presented in this chapter,

$$Arg[C(j\omega_{pc})P(j\omega_{pc})] = \pi. \tag{4.2}$$

iv. Gain margin (A_m), an explicit indicator of stability,

$$A_m = -20lg|C(j\omega_{pc})P(j\omega_{pc})|. \tag{4.3}$$

v. 'Flat phase constraint at ω_{gc}, which means the system robustness to gain variations and is expressed as (3.13).

The closed-loop transfer function in Figure 4.1 is

$$G(s) = \frac{C(s)P(s)M_T(A,\phi)}{1+C(s)P(s)M_T(A,\phi)}. \tag{4.4}$$

The characteristic equation is $D(s) = 1+C(s)P(s)M_T(A,\phi)$. Substituting (4.1) (3.21) (2.3) into (4.4), yields

$$D(s) = s^\lambda(s+1)+Ae^{-j\phi}e^{-Ls}(k_p s^\lambda + k_i + k_d s^{\lambda+\mu}). \tag{4.5}$$

Remark 1. *The complex root boundary (CRB) is defined as $D(s = j\omega) = 0$ in (4.5) [29, 32].*

Substitute $s = jw$ in (4.5), and the CRB can be obtained as follows:

$$D(s=j\omega) = (j\omega)^\lambda(j\omega+1)+Ae^{-j(\phi+\omega L)}(k_p(j\omega)^\lambda + k_i + k_d(j\omega)^{\lambda+\mu}) = 0. \tag{4.6}$$

Then, equating the real and imaginary items of (4.6) to 0, respectively, yields

$$B_1 + AC_1E + AS_1F = 0, \tag{4.7}$$
$$B_2 + AC_1F - AS_1E = 0, \tag{4.8}$$

where

$$\begin{aligned}
&B_1 = \omega^\lambda C_2 - \omega^{1+\lambda}S_2, \quad B_2 = \omega^{\lambda+1}C_2 + \omega^\lambda S_2,\\
&C_1 = cos(\phi+\omega L), \quad C_2 = cos\frac{\lambda\pi}{2}, \quad C_3 = cos\frac{(\lambda+\mu)\pi}{2},\\
&S_1 = sin(\phi+\omega L), \quad S_2 = sin\frac{\lambda\pi}{2}, \quad S_3 = sin\frac{(\lambda+\mu)\pi}{2},\\
&E = k_p\omega^\lambda C_2 + k_i + k_d\omega^{\mu+\lambda}C_3, \quad F = k_p\omega^r S_2 + k_d\omega^{\lambda+\mu}S_3.
\end{aligned}$$

Solving the equation set (4.7) and (4.8), k_p and k_i can be obtained:

$$k_p = -\frac{B_1S_1+B_2C_1}{A\omega^\lambda S_2} - \frac{k_d\omega^\mu S_3}{S_2}, \tag{4.9}$$

$$k_i = \frac{B_2 - B_1C_1S_1 - B_2C_1^2}{AS_1} + \frac{B_1S_1C_2+B_2C_1C_2}{AS_2} + \frac{k_d\omega^{\lambda+\mu}S_3C_2}{S_2} - k_d\omega^{\lambda+\mu}C_3. \tag{4.10}$$

Given a fixed k_d, λ, and μ, a line of k_p versus k_i can be plotted through above equations with ω sweeping from $0 \to +\infty$.

Remark 2. *The real root boundary (RRB) is defined as $D(s=0)=0$ in (4.5), so the result is $k_i=0$.*

Remark 3. *The stable boundaries of the FOPI$^\lambda$D$^\mu$ parameters can be determined by the CRB and RRB defined in Remark 2 and Remark 3, respectively, and the stable region can be determined through random point testing [29].*

Considering the phase margin ϕ_m specification, setting $\phi = \phi_m$ and $A = 1$ in (4.9) and (4.10), one line l can be obtained in the parameter space with a fixed k_d, λ, and μ, and increasing ω from 0. All the points on the line meet the phase margin ϕ_m [29].

Considering a given ω_{gc} specification, a point corresponding to the parameters (k_p, k_i) on the above line l can be determined with $\omega = \omega_{gc}$ [33]. Sweeping all the $\lambda \in (0,2)$, a set of points satisfying the above two frequency-domain specifications ω_{gc} and ϕ_m can be obtained.

Considering the flat phase constraint, from (4.7) and (4.8), one can get

$$S_1 = \frac{B_2E - B_1F}{A(E^2+F^2)}, \tag{4.11}$$

$$C_1 = \frac{-B_1E - B_2F}{A(E^2+F^2)}. \tag{4.12}$$

so,

$$\phi = arctan\frac{B_1F - B_2E}{B_1E + B_2F} - \omega L. \tag{4.13}$$

The flat phase constraint means the phase plot is flat. From (4.13),

$$\frac{\mathrm{d}\phi}{\mathrm{d}\omega} = \frac{(B_1^2+B_2^2)(EF'-E'F)+(B_1{}'B_2 - B_1B_2{}')(E^2+F^2)}{(B_1E+B_2F)^2+(B_1F-B_2E)^2} - L = 0. \tag{4.14}$$

where

$$\begin{aligned}
E' &= \lambda k_p\omega^{\lambda-1}C_2 + (\mu+\lambda)k_d\omega^{\lambda+\mu-1}C_3,\\
F' &= \lambda k_p\omega^{\lambda-1}S_2 + (\mu+\lambda)k_d\omega^{\lambda+\mu-1}S_3,\\
B_1' &= \lambda\omega^{\lambda-1}C_2 - (1+\lambda)\omega^{\lambda}S_2,\\
B_2' &= (\lambda+1)\omega^{\lambda}C_2 + \lambda\omega^{\lambda-1}S_2.
\end{aligned}$$

All the points satisfying both the frequency-domain specifications ω_{gc} and ϕ_m can be checked by Eq. (4.14). Assuming that a certain point with the FOPI$^\lambda$D$^\mu$ controller parameters can be found as a solution of Eq. (4.14), it is the flat phase point. Sweeping all the $\mu \in (0,2)$, a series of flat phase points will be obtained. All the points satisfy three frequency-domain specifications: ω_{gc}, ϕ_m, and the flat phase constraint.

Considering the phase crossover frequency ω_{pc} and gain margin A_m, the phase and gain of the $C(s)$ are

$$\mathrm{Arg}[C(j\omega)] = \tan^{-1}\frac{D_1}{D_2}, \tag{4.15}$$

$$|C(j\omega)| = \sqrt{D_1^2 + D_2^2}, \tag{4.16}$$

where

$$D_1 = -k_i\omega^{-\lambda} sin\frac{\lambda\pi}{2} + k_d\omega^{\mu} sin\frac{\mu\pi}{2},$$

$$D_2 = k_p + k_i\omega^{-\lambda} cos\frac{\lambda\pi}{2} + k_d\omega^{\mu} cos\frac{\mu\pi}{2}.$$

The phase and gain of $P(s)$ are given as

$$\mathrm{Arg}[P(j\omega)] = -\tan^{-1}(\omega) - \omega L, \tag{4.17}$$

$$|P(j\omega)| = \frac{1}{\sqrt{1+(\omega)^2}}. \tag{4.18}$$

The phase and gain of the open-loop transfer function are

$$\mathrm{Arg}[C(j\omega)P(j\omega)] = \tan^{-1}\frac{D_1}{D_2} - \tan^{-1}(\omega) - \omega L, \tag{4.19}$$

$$|C(j\omega)P(j\omega)| = \frac{\sqrt{D_1^2 + D_2^2}}{\sqrt{1+(\omega)^2}}. \tag{4.20}$$

According to the definition of phase crossover frequency, one can get

$$\mathrm{Arg}[C(j\omega_{pc})P(j\omega_{pc})] = \tan^{-1}\frac{D_1}{D_2} - \tan^{-1}(\omega_{pc}) - \omega_{pc}L = -\pi. \tag{4.21}$$

Similarly, according to the definition of gain margin, one can get

$$|C(j\omega_{pc})P(j\omega_{pc})|_{dB} = -\log_{10}\left(\frac{\sqrt{D_1^2 + D_2^2}}{\sqrt{1+(\omega_{pc})^2}}\right) = A_m. \tag{4.22}$$

All the flat phase points can be detected using Eq. (4.21). Assuming that a point with the FOPI$^\lambda$D$^\mu$ controller parameters can be found as a solution of Eq. (4.21), this flat phase point satisfies the frequency-domain specification phase crossover frequency. Sweeping all the K_d in the reasonable range $(-T, T)$ [69], one can obtain a series of points satisfying four frequency-domain specifications. Then, testing all the points by Eq. (4.22), the solution corresponding to the FOPI$^\lambda$D$^\mu$ is the designed controller, and the control system satisfies five frequency-domain specifications: ω_{gc}, ϕ_m, ω_{pc}, A_m, and the flat phase constraint.

4.3 DESIGN PROCEDURE ILLUSTRATION WITH AN EXAMPLE

Step 1. Considering the FOPNTD system, given the frequency-domain specification phase margin, $\phi_m = 50^\circ$ and $L = 0.1$ s.

Step 2. Given a fixed $\lambda = 0.8/1.0/1.2$, $k_d = 0.1$ and $\mu = 0.9$, the lines of k_p versus k_i can be plotted as shown in Figure 4.2 as ω increasing from 0. All the points on the lines satisfy $\phi_m = 50^\circ$.

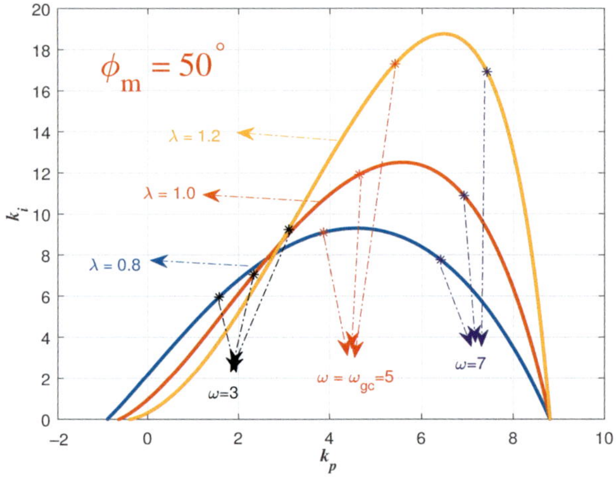

FIGURE 4.2 $\phi_m = 50^\circ$ with $k_d = 0.1$, $\mu = 0.9$, and different λ.

Step 3. Given the frequency-domain specification gain crossover frequency $\omega_{gc} = 5$, the point (k_p, k_i) can be fixed as red stars in Figure 4.2. Sweeping over $\lambda \in (0,2)$, curves can be drawn in Figure 4.3 with fixed $\mu = 0.6/1.0/1.4$ in the three-dimension parameter space (k_p, k_i, λ). All points on the above curves meet $\phi_m = 50^\circ$ and $\omega_{gc} = 5$.

Step 4. Select all points on above curves to find the solutions of (4.14). The solutions corresponding the points are shown as squares in Figure 4.3, which are the flat phase points. These points corresponding the controller parameters (k_p, k_i, k_d, λ, μ) satisfy ϕ_m, ω_{gc}, and the flat phase.

Step 5. Given the frequency-domain specification $\omega_{pc} = 18$ rad/s, sweeping over the $\mu \in (0,2)$, all flat phase points can be obtained as shown in Figure 4.4, and the Bode plots corresponding to the points are shown in Figure 4.5. Test all the points in Figure 4.4 to find the solution of Eq. (4.21). The solution point is shown as the red circle in Figure 4.4. This point is corresponding to the controller with parameters satisfying, ϕ_m, ω_{gc}, the flat phase, and ω_{pc}. The open-loop Bode plot associated the with this point is shown as the red line in Figure 4.5.

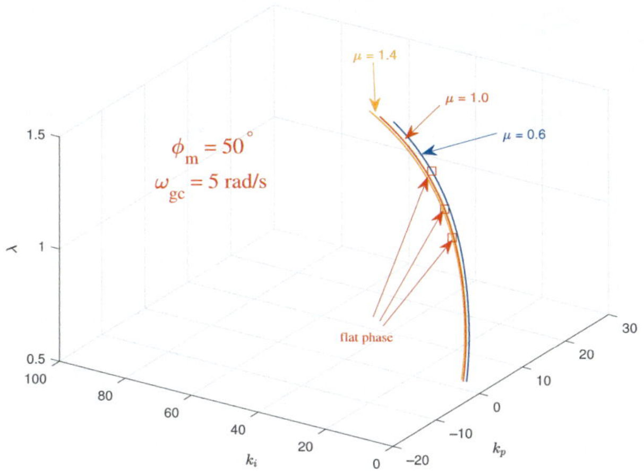

FIGURE 4.3 $\phi_m = 50^\circ$ and $\omega_{gc} = 5$ rad/s.

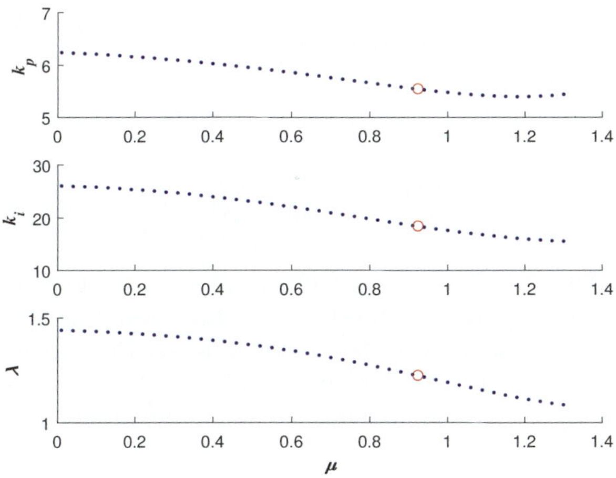

FIGURE 4.4 The points satisfy ϕ_m, ω_{gc}, and the flat phase.

The complementary sensitivity function (the closed-loop transfer function), which reflects the step-tracking performance with overshoot level, is

$$T(s) = \frac{C(s)P(s)}{1+C(s)P(s)}, \tag{4.23}$$

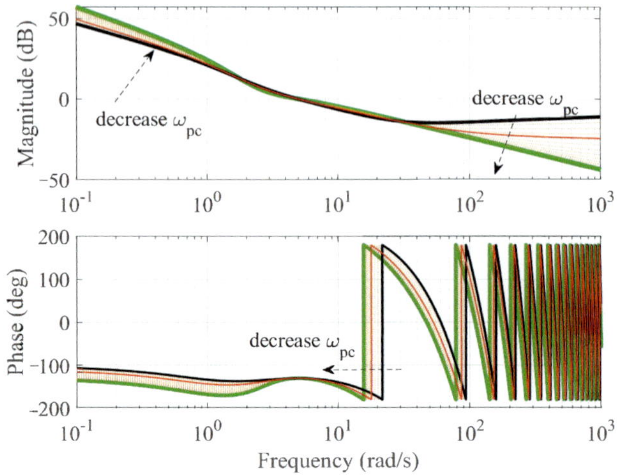

FIGURE 4.5 Bode plots of flat phase points.

the sensitivity function, which reflects the rejection of the output disturbances, is

$$S(s) = \frac{1}{1 + C(s)P(s)}, \tag{4.24}$$

and the process sensitivity function, which reflects process disturbances rejection, is

$$E(s) = \frac{P(s)}{1 + C(s)P(s)}. \tag{4.25}$$

The Bode plots of the function $T(s)$ of all the flat phase points in Figure 4.4 are presented in Figure 4.6a. Figure 4.6a shows that the resonance peak decreases as ω_{pc} increases, which means that the overshoots will decrease. The step responses, simulations are carried out and shown in Figure 4.6b. The results show that the overshoots decrease with the ω_{pc} increasing, which is the verification of the frequency-domain analysis.

The Bode plots of the functions $S(s)$ of all the flat phase points in Figure 4.4 are shown in Figure 4.7a. Figure 4.7a shows that the ability of low-frequency disturbance rejection increases as ω_{pc} decreases, and it can be seen from Figures 4.6a and 4.7a that the ability of high-frequency disturbance rejection increases as ω_{pc} decreases. Setting the reference signal 0, inputting low-frequency disturbance and high-frequency noise, the simulations are carried out and the results are presented in Figure 4.7b. Obviously, the amplitudes of the responses decrease as ω_{pc} decreases. For more clearly, the fast Fourier transform (FFT) of the responses with low-frequency disturbance and high-frequency noise is calculated and presented in Figure 4.8. According to the Figure 4.8, the magnitude of peak at 0.1 Hz decreases when the ω_{pc}

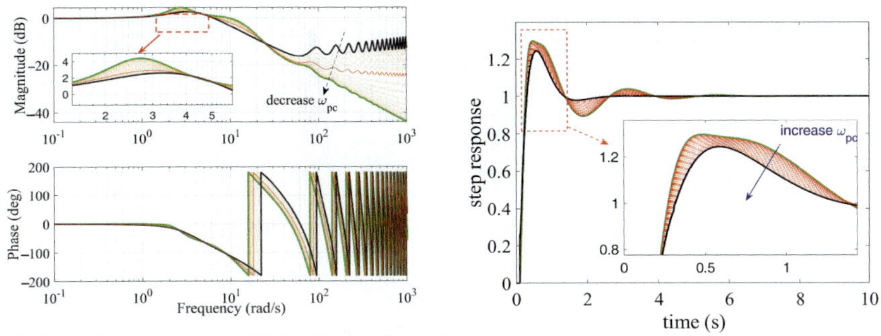

(a) Complementary sensitivity Bode plots of flat phase points.

(b) Step response.

FIGURE 4.6 Complementary sensitivity Bode plots and step responses comparison.

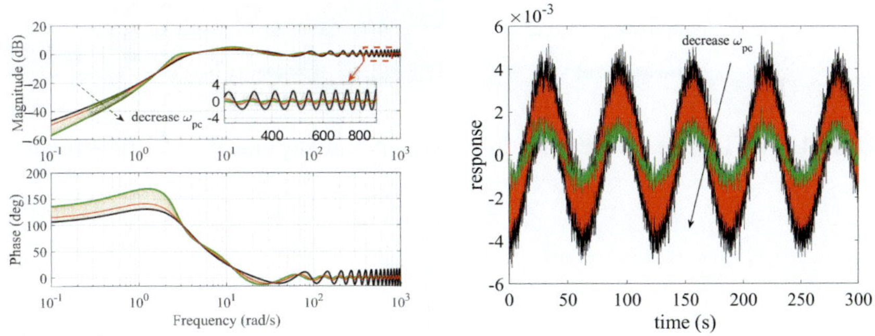

(a) Sensitivity Bode plots.

(b) Responses.

FIGURE 4.7 Sensitivity Bode plots of flat phase points and responses with low-frequency disturbance and high-frequency noise.

decreases. Moreover, the magnitude of peak at high frequencies also decreases when the ω_{pc} decreases. The results of time-domain responses and FFT are consistent with frequency-domain analysis.

The Bode plots of function $E(s)$ of all the flat phase point in Figure 4.4 are presented in Figure 4.9a, which demonstrates that the peak decreases with the ω_{pc} is increasing, which means the ability of process disturbance rejection increases as ω_{pc} increasing. Giving the same step disturbance, the simulation results are presented in Figure 4.9b. When increasing ω_{pc}, the response has less oscillation and takes shorter time to recover to the reference value. That means the process disturbance rejection capability increases as ω_{pc} increases, which is a solid validation for the frequency-domain analysis.

Step 6. Given the frequency-domain specification $A_m = 10$ dB, sweeping all the $k_d \in (-1, 1)$, all the points satisfying four frequency-domain specifications can be

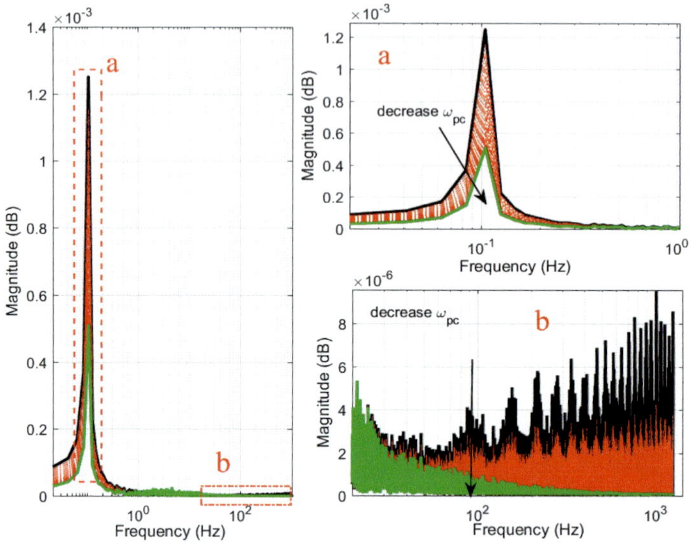

FIGURE 4.8 FFT of response with low-frequency disturbance and high-frequency noise.

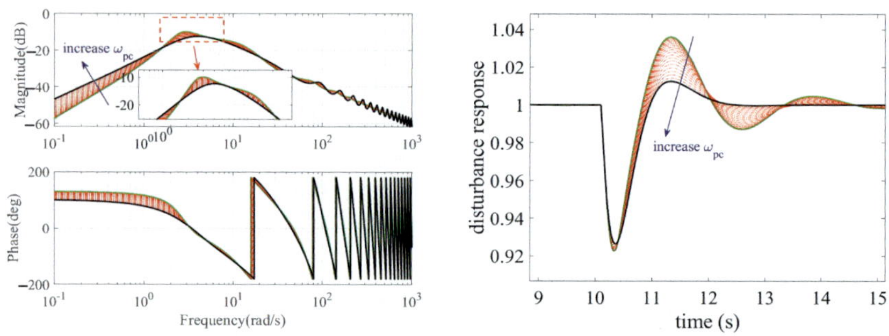

(a) Process sensitivity Bode plots.

(b) Responses with step disturbance.

FIGURE 4.9 Process sensitivity Bode plots of flat phase points and responses with step disturbance.

obtained as shown in Figure 4.10. Selecting all the points corresponding controller parameters and taking into account Eq. (4.22) for verification, the solution can be determined and shown as a black triangle in Figure 4.10. The corresponding Bode plot is presented in Figure 4.11.

The pseudo code of the synthesis implementation for the FOPI$^{\lambda}$D$^{\mu}$ controller analytical design is shown in Algorithm 1. Note that the design synthesis procedure

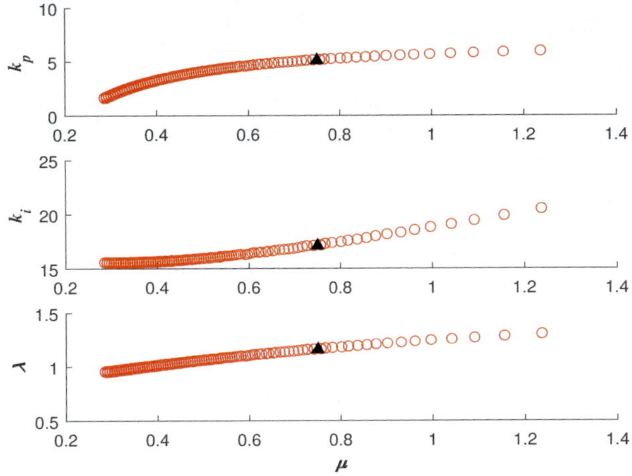

FIGURE 4.10 The points satisfy ϕ_m, ω_{gc}, the flat phase, and A_m.

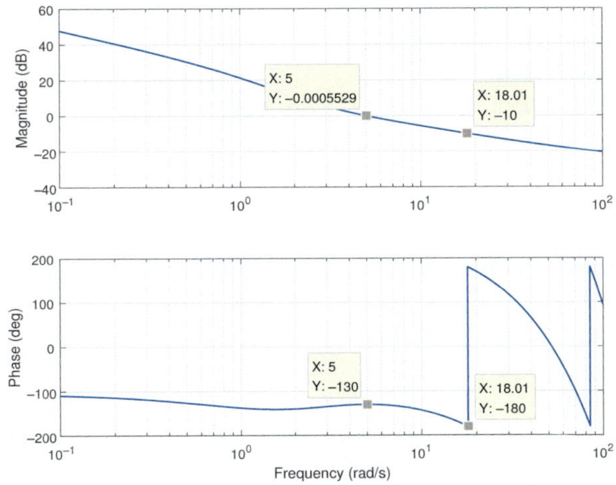

FIGURE 4.11 Bode plot of the designed FOPI$^\lambda$D$^\mu$ control system.

is also applicable to the FOPTD system. Furthermore, the code is applicable to all the FOPID-type controllers with some of the five specifications. For example, FOPI with three frequency-domain specifications [33], IOPID with three frequency-domain specifications [69], FOPD with three frequency-domain [15], etc.

Algorithm 1 The pseudo code of the synthesis implementation for FOPI$^{\lambda}$D$^{\mu}$ controller analytical design.

for $k_d = -1 : small\ interval : 1$ **do**
  for $\mu = 0 : small\ interval : 1$ **do**
    $\omega = \omega_{gc}$, $\phi = \phi_m$; %Given the frequency-domain specifications
    $r = 0 : small\ interval : 2$;
    $k_p = Eq.\ (4.9)$;
    $k_i = Eq.\ (4.10)$;
    $\frac{\mathrm{d}\phi}{\mathrm{d}\omega} = Eq.\ (4.14)$; %The flat phase constraint
    for $j = 1 : 1 : (2-0)/small\ interval$ **do**
      $Der = \frac{\mathrm{d}\phi}{\mathrm{d}\omega}(1, j) * \frac{\mathrm{d}\phi}{\mathrm{d}\omega}(1, j+1)$;
      if $Der < 0$ **then**
        The FOPI$^{\lambda}$D$^{\mu}$ controller parameters satisfying ω_{gc}, ϕ_m and flat phase constraint can be obtained;
      end if
    end for
  end for
  $\omega = \omega_{pc}$; %Given the frequency-domain specifications
  $arg = Eq.\ (4.21)$;
  for $i = 1 : 1 : length(arg) - 1$ **do**
    $arg = arg(1,i) * arg(1,i+1)$;
    if $arg < 0$ **then**
      The FOPI$^{\lambda}$D$^{\mu}$ controller parameters satisfying ω_{gc}, ϕ_m, ω_{pc} and flat phase constraint can be obtained;
    end if
  end for
end for
$A = A_m$; %Given the frequency-domain specifications
$|A| = Eq.\ (4.22)$;
for $k = 1 : 1 : length(|A|) - 1$ **do**
  $|A| = |A|(1,k) * |A|(1,k+1)$;
  if $|A| < 0$ **then**
    The FOPI$^{\lambda}$D$^{\mu}$ controller parameters satisfying ω_{gc}, ϕ_m, ω_{pc}, A_m and flat phase constraint can be determined;
  end if
end for

4.4 ACHIEVABLE REGION COLLECTION GUIDELINE

The analytical design for FOPI$^{\lambda}$D$^{\mu}$ controller has been summarized. However, how do we know the chosen frequency-domain specifications are available to obtain a stabilizing controller? Thus, in this section, all achievable frequency-domain specifications, i.e. ω_{gc}, ϕ_m, ω_{pc}, and A_m can be visualized in multidimensional graphics, which is named the achievable region plot. This achievable region gives users prior information and great flexibility before the controller design.

Given a fixed ϕ_m and ω_{gc}, all points satisfying the flat phase can be collected and tested by Eq. (4.21) with phase crossover frequency $\omega_{pc} \in (\omega_{gc}, \omega_{pcmax})$. Thus, the achievable ω_{pc} can be determined and the corresponding controller parameters can be collected. Next, the achievable gain margin can be calculated by Eq. (4.22). Sweeping the $\omega_{gc} \in (0, \omega_{gcmax})$ with the fixed ϕ_m, the achievable ω_{gc}, ω_{pc}, and A_m can be collected. Finally, sweeping all $\phi_m \in (0, 180^\circ)$, the complete achievable region for ω_{gc}, ϕ_m, ω_{pc}, and A_m, guaranteeing the FOPI$^{\lambda}$D$^{\mu}$ satisfying the flat phase, can be collected. The pseudo code for collecting the achievable region of frequency-domain specifications is shown in Algorithm 2.

Algorithm 2 The pseudo code for collecting the achievable region of frequency-domain specifications.

```
for φ = 0 : 1 : 180 do
    for ω = 0 : small interval : ω_gcmax do
        for k_d = −1 : small interval : 1 do
            for μ = 0 : small interval : 1 do
                r = 0 : small interval : 2;
                k_p = Eq. (4.9);
                k_i = Eq. (4.10);
                dφ/dω = Eq. (4.14);  %The flat phase constraint
                for j = 1 : 1 : (2 − 0)/small interval do
                    Der = dφ/dω(1, j) * dφ/dω(1, j + 1);
                    if Der < 0 then
                        The FOPI^λ D^μ controller parameters satisfying ω_gc, φ_m
                        and flat phase constraint can be obtained;
                    end if
                end for
            end for
        end for
        for ω_pc = w : small interval : ω_pcmax do
            arg = Eq. (4.21);
            for i = 1 : 1 : length(arg) − 1 do
                arg = arg(1, i) * arg(1, i + 1);
                if arg < 0 then
                    The FOPI^λ D^μ controller parameters satisfying ω_gc, φ_m, ω_pc
                    and flat phase constraint can be obtained;
                end if
            end for
        end for
        A_m = Eq. 4.22
    end for
end for
```

Thus, given $L = 0.1$, the achievable region can be collected and visualized as shown in Figure 4.12, which includes all the achievable specifications: ω_{gc}, ϕ_m, ω_{pc}, and A_m.

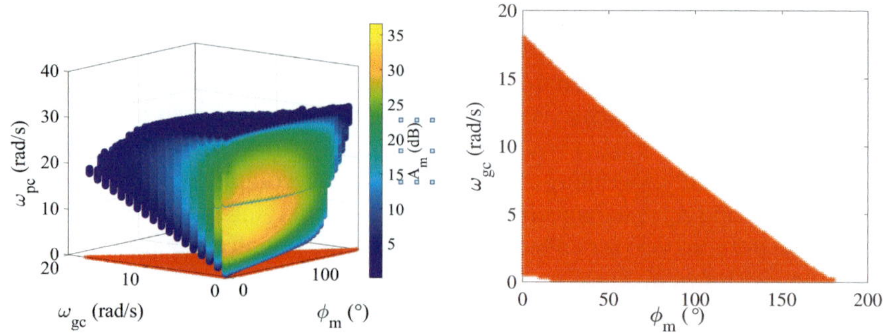

(a) Complete achievable region. **(b)** Achievable region of ϕ_m and ω_{gc}.

FIGURE 4.12 Achievable region with $L = 0.1$.

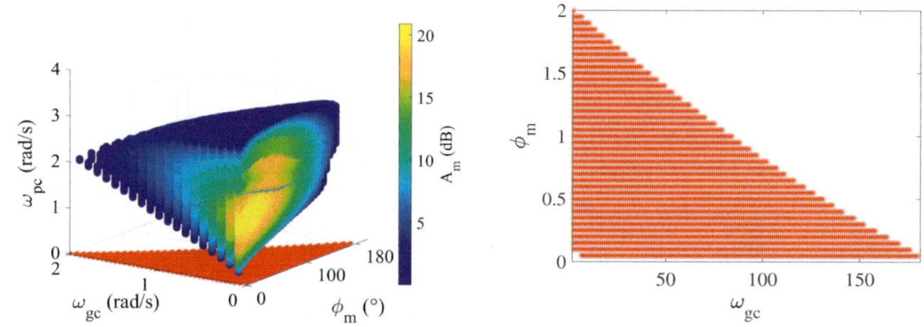

(a) Complete achievable region. **(b)** Achievable region of ϕ_m and ω_{gc}.

FIGURE 4.13 Achievable region with $L = 1$.

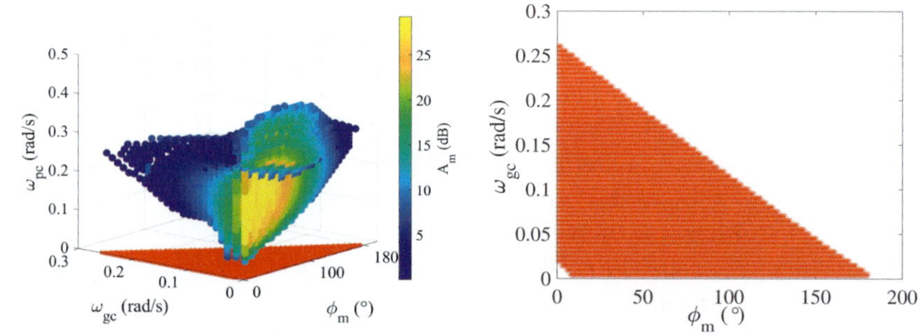

(a) Complete achievable region. **(b)** Achievable region of ϕ_m and ω_{gc}.

FIGURE 4.14 Achievable region with $L = 10$.

The red area expresses the achievable gain crossover frequency and phase margin, which can be presented in Figure 4.12a. For the FOPNTD systems, the complete achievable regions can be obtained for $L = 1$ and $L = 10$ in Figures 4.13 and 4.14, respectively.

4.5 SIMULATION RESULTS

Several illustrative simulation examples are carried out in this section to verify the effectiveness of the proposed design synthesis of the FOPI$^\lambda$D$^\mu$ controller for different FOPTD systems, which is compared with other typical controllers and optimal design methods in Chen et al. [34], Chevalier et al. [24], Yumuk et al. [47] and Valrio et al. [51]. The rise time (t_r/s), settling time (t_s/s), overshoot (σ/%), speed drop (κ/%), and ITAE are used to compare the control performance. The fractional order (FO) operators s^λ are implemented using impulse response invariant discretization methods [70].

4.5.1 Example 1: Lag-Dominated Process

The previously studied lag-dominated process is considered:

$$G(s) = \frac{1}{(s+1)(0.2s+1)}. \tag{4.26}$$

It can be approximated to the FOPTD by

$$G_1(s) = \frac{1}{1.11s+1}e^{-0.105s}. \tag{4.27}$$

Given the frequency-domain specifications $\omega_{gc} = 2$ rad/s and $\phi_m = 60^\circ$, various controllers can be obtained using the design methods in [24, 34, 47, 51] as

$$C_{Proposed1}(s) = 2.4142 + \frac{3.3732}{s^{1.11}} - 0.315s^{0.175}, \tag{4.28}$$

$$C_{Chen}(s) = 1.8032 + \frac{3.5521}{s} + 0.07s, \tag{4.29}$$

$$C_{Chevalier}(s) = 1.5432(1 + \frac{2.4041}{s^{0.9265}} + 0.104 + s^{0.9265}), \tag{4.30}$$

$$C_{Yumuk}(s) = 2.5495s^{-0.1996}(1 + \frac{0.9009}{s}), \tag{4.31}$$

$$C_{Valrio}(s) = 0.3669(1 + \frac{0.4499}{s^{1.4191}} - 0.0905s^{0.9981}). \tag{4.32}$$

Figure 4.15a illustrates the open-loop Bode diagrams of five control systems. The step-tracking and anti-load disturbance responses of the designed controllers in (4.28)-(4.29) are depicted in Figure 4.15b. The frequency- and time-domain responses of five control systems are summarized in Tables 4.1 and 4.2, respectively.

According to the Table 4.1 and Figure 4.15a, all the control systems satisfy the given frequency-domain specifications expected for that of Valrio et al., in which the

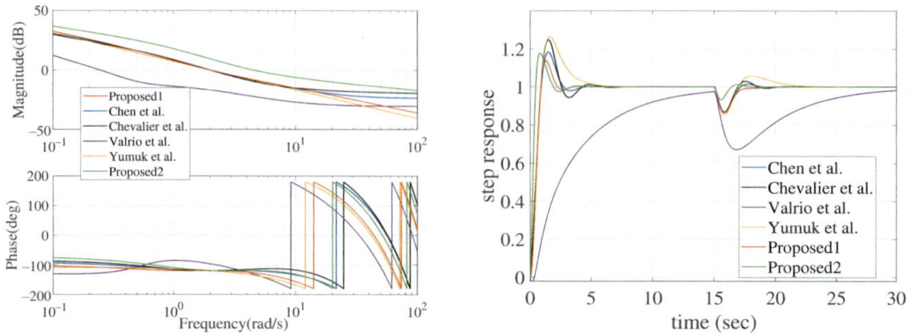

(a) Open-loop Bode plots. **(b)** Step-tracking responses.

FIGURE 4.15 Open-loop Bode plots, step-tracking, and anti-load disturbance responses of lag-dominated control systems.

TABLE 4.1
The Frequency-Domain Specifications of Lag-Dominated Control Systems

Controllers	ω_{gc} (rad/s)	$\phi_m (^\circ)$	ω_{pc} (rad/s)	A_m (dB)
Proposed 1	2	60	14	18.15
Chen et al.	2	60	21.42	20.51
Chevalier et al.	2	60	24.72	17.89
Yumuk et al.	2	60	11.97	18.65
Valrio et al.	0.2455	52.1	9.161	26.64
Proposed 2	4.5	60	20	10

TABLE 4.2
The Time-Domain Performance of Lag-Dominated Control Systems

	Step Tracking				Anti-Load Disturbance	
Controller	Overshoot (%)	Rise Time (s)	Settling Time (s)	$ITAE_{st}$	Speed Drop (%)	$ITAE_{ld}$
Proposed 1	13.8	0.8655	2.05	0.3264	13.54	2.8626
Chen et al.	18.4	0.899	3.398	0.5235	13.41	2.7535
Chevalier et al.	24.8	0.8815	3.912	0.8154	13.23	2.9996
Valrio et al.	0	15	15	13.41	32.87	38.36
Yumuk et al.	26.4	0.897	4.685	1.294	14.61	5.638
Proposed 2	17.8	0.44	2.721	0.2518	6.77	1.0412

controller design is used as the *second set of the formula* in [51]. Figure 4.15b and Table 4.2 show that the control system of controller (4.28) has the shortest settling time (2.05 s). Although the control system of Valrio et al. has no overshoot, the settling time is the longest (15 s). The values of the $ITAE_{st}$ show that the proposed controller (4.28) achieves the best step-tracking performance among all five control systems. Furthermore, the proposed controller (4.28) from Chen et al. and Chevalier et al. has a better anti-load disturbance performance than others. In summary, the proposed controller (4.28) achieves the best control performance from the sum value of J_{ITAE} of the step-tracking and anti-load disturbance.

To demonstrate the flexibility of the proposed design approach of FOPI$^{\lambda}$D$^{\mu}$, one more controller is designed with the same phase margin and bigger gain crossover frequency $\omega_{gc} = 4.5$ rad/s. The designed FOPI$^{\lambda}$D$^{\mu}$ is

$$C_{Proposed2}(s) = 2.4631 + \frac{10.7462}{s^{0.8}} + s^{0.565}. \tag{4.33}$$

The open-loop Bode plot and step response are shown in Figure 4.15a and 4.15b, respectively. The frequency- and time-domain characteristics are summarized in Tables 4.1 and 4.2, respectively.

According to the Figure 4.15b and Table 4.2, the designed controller (4.33) achieves bigger overshoot and settling time than the controller (4.28) but the shorter rise time (0.44 s). Obviously, the speed drop is only half of that of (4.28) (4.29) (4.30). In summary, the designed controller (4.33) achieves the better step tracking and especially anti-load disturbance performance.

4.5.2 Example 2: Lag-Delay Balanced Process

Consider the lag-delay balanced plant studied in [71]:

$$G(s) = \frac{1}{s+1}e^{-0.67s}. \tag{4.34}$$

Various controllers can be designed with the given $\omega_{gc} = 0.4$ and $\phi_m = 60^\circ$ (note that the controller of Chen et al. cannot be designed with above frequency-domain indicators according to the proposed design method in [34]):

$$C_{Proposed1}(s) = 0.2257 + \frac{0.4306}{s^{1.17}} + 0.3850s^{0.345}, \tag{4.35}$$

$$C_{Chevalier}(s) = 0.5723(1 + \frac{0.6868}{s^{1.2235}} + 0.364 + s^{1.2235}), \tag{4.36}$$

$$C_{Yumuk}(s) = 0.34462s^{-0.16272}(1 + \frac{1}{s}), \tag{4.37}$$

$$C_{Valrio}(s) = 0.587(1 + \frac{0.589}{s^{1.346}} + 0.2168s^{1.0899}). \tag{4.38}$$

Figure 4.16a illustrates the open-loop diagram plots of four control systems. The step-tracking and anti-load disturbance responses of various controllers in

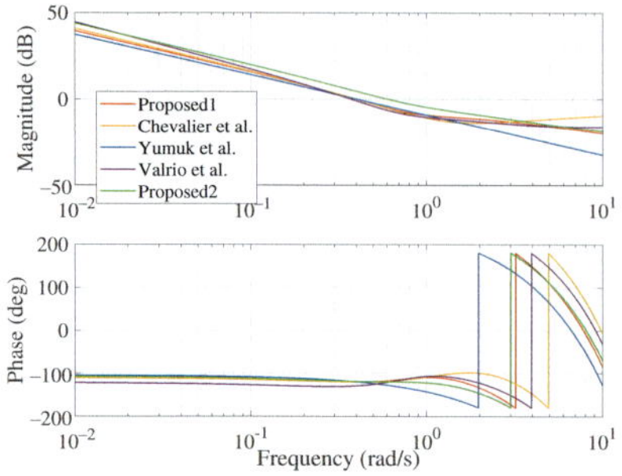

(a) Open-loop Bode plots.

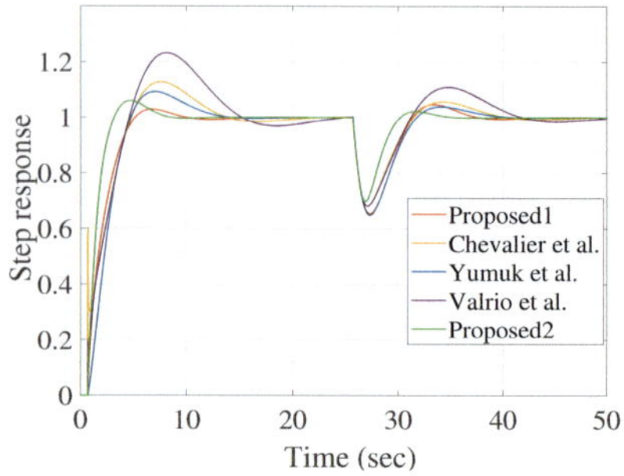

(b) Step-tracking responses.

FIGURE 4.16 Open-loop Bode plots and step-tracking, anti-load disturbance responses of lag-delay balanced control systems.

(4.35)–(4.38) are depicted in Figure 4.16b. The frequency-domain and time-domain characteristics of four control systems are summarized in Tables 4.3 and 4.4, respectively.

TABLE 4.3
The Frequency-Domain Specifications of Lag-Delay Balanced Control Systems

Controllers	ω_{gc} **(rad/s)**	$\phi_m(^\circ)$	ω_{pc} **(rad/s)**	A_m **(dB)**
Proposed 1	0.4	60	3.2	13.4
Chen et al.		Unobtainable		
Chevalier et al.	0.4	60	4.933	11.58
Yumuk et al.	0.4	60	1.963	16.6
Valrio et al.	0.375	50.3	3.942	15.24
Proposed 2	0.6	60	3	11.8

TABLE 4.4
The Time-Domain Performance of Lag-Delay Balanced Control Systems

	Step Tracking				**Anti-Load Disturbance**	
Controller	**Overshoot (%)**	**Rise Time (s)**	**Settling Time (s)**	$ITAE_{st}$	**Speed Drop (%)**	$ITAE_{ld}$
Proposed 1	2.9	5.135	8.05	3.663	34.59	38.097
Chevalier et al.	12.8	4.762	12.55	9.894	31.94	39.466
Yumuk et al.	9.2	4.71	11.64	7.17	35.15	38.96
Valrio et al.	23.2	4.59	21.11	19.7	31.83	54.9
Proposed 2	6.0	3.165	6.924	2.225	30.2	21.005

For the lag-balanced process (4.34), the controller of Chen et al. cannot be designed with the given specifications, and other designed control systems satisfy the given specifications except for the control system of Valrio et al., for which the controller design is based on *a second set of the formula*. Figure 4.16b and Table 4.4 show that the proposed controller (4.35) has the smallest overshoot (2.9%) and settling time (8.05 s). The values of the $ITAE_{st}$ show that the proposed controller achieves the best step-tracking performance. Furthermore, the controllers of the proposed (4.35) and Valrio et al. achieve similar anti-load disturbance performance, and are both superior to other controllers. In summary, the proposed controller achieves better performance than other controllers for the lag-delay balanced process.

To demonstrate the flexibility of the proposed design approach of the FOPI$^\lambda$D$^\mu$, one more controller is designed with the same ϕ_m and bigger gain crossover frequency $\omega_{gc} = 0.6$. The designed FOPI$^\lambda$D$^\mu$ controller is

$$C_{Proposed2}(s) = 0.8099 + \frac{0.6426}{s^{1.19}} + 0.1s^{0.885}. \tag{4.39}$$

According to the Figure 4.16b and Table 4.4, the new designed FOPI$^\lambda$D$^\mu$ controller (4.39) achieves the shortest rise time (3.165 s) and settling time (6.924 s). Furthermore, the performance of anti-load disturbance has been significantly improved.

4.5.3 Example 3: Delay-Dominated Process

The following delay-dominated plant is considered:

$$G(s) = \frac{1}{s+1}e^{-1.5s}. \tag{4.40}$$

Given the frequency-domain specifications $\omega_{gc} = 0.45$ rad/s and $\phi_m = 40°$, the controllers can be designed as

$$C_{Proposed}(s) = 0.82929 + \frac{0.4546}{s^{1.31}} + 0.15s^{0.825}, \tag{4.41}$$

$$C_{Yumuk}(s) = 0.407s^{0.1258}(1 + \frac{1}{s}). \tag{4.42}$$

Note that the controllers of Chen et al. and Chevalier et al. cannot be designed with the given specifications, and the controller of Valrio et al. is designed based on *second set of the formula*.

Figure 4.17a illustrates the open-loop Bode plots. The step-tracking and anti-load disturbance responses of various controllers in (4.41)–(4.42) are depicted in Figure 4.17b. The frequency-domain and time-domain characteristics of three control systems are summarized in Tables 4.5 and 4.6, respectively.

For the delay-dominated process (4.40), the controllers of Chen et al. and Chevalier et al. cannot be designed with the given specifications. Other controllers can be

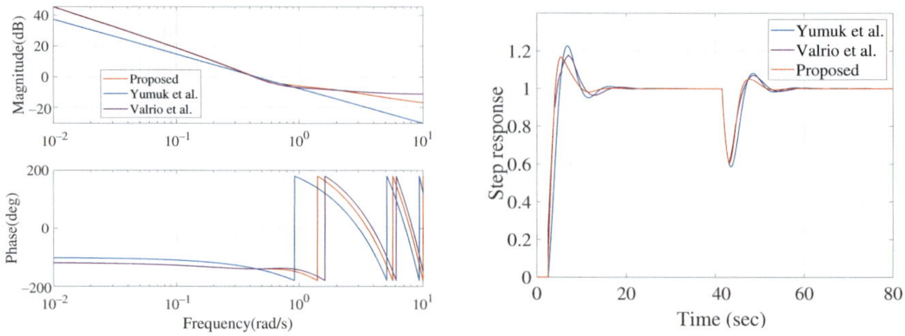

(a) Open-loop Bode plots. **(b)** Step-tracking responses.

FIGURE 4.17 Open-loop Bode plots and step-tracking, anti-load disturbance responses of delay-dominated control systems.

obtained with the given specifications. The proposed FOPI$^{\lambda}$D$^{\mu}$ controller has the smallest $ITAE_{st}$ with the smallest overshoot (16.8%), rise time (4.076 s) and settling time (s). Moreover, in terms of anti-load disturbance performance, the controller of Yumuk et al. achieves the smallest speed drop (38.9%), but has a longer recovery time compared to the proposed controller according to Figure 4.17b. Thus, the proposed

TABLE 4.5
The Frequency-Domain Specifications of Delay-Dominated Control Systems

Controllers	ω_{gc} (rad/s)	$\phi_m (°)$	ω_{pc} (rad/s)	A_m (dB)
Proposed	0.45	40	1.4	7.2
Chen et al.		Unobtainable		
Chevalier et al.		Unobtainable		
Yumuk et al.	0.45	40	0.9153	6.943
Valrio et al.	0.4309	39.9	1.611	8.249

TABLE 4.6
The Time-Domain Performance of Delay-Dominated Control Systems

	Step Tracking				Anti-Load Disturbance	
Controller	Overshoot (%)	Rise time (s)	Settling time (s)	$ITAE_{st}$	Speed drop (%)	$ITAE_{ld}$
Proposed	16.8	4.076	9.18	8.586	39.35	52.424
Yumuk et al.	22.6	4.497	14.66	13.73	38.9	72.1
Valrio et al.	18.0	5.061	13.38	11.23	41.48	58.92

FOPI$^{\lambda}$D$^{\mu}$ controller achieves the best control performance for the delay-dominated process from the values of $ITAE_{st}$ and $ITAE_{ld}$.

4.5.4 Summary and Further Discussion

In this subsection, the comprehensive performance of five controllers for lag-dominated, lag-delay balanced and delay-dominated processes is summarized in Table 4.7. From Table 4.7, the proposed controllers, from Valrio et al. and Yumuk et al., are applicable to a wider range of FOPTD process than other controllers. That means the achievable region of frequency-domain specifications of the proposed controller is bigger than those of the controllers in Chen [34] and Chevalier [24]. Furthermore, the proposed controller achieves the best control performance over the other controllers for all types processes according to the values of $ITAE_{sum}$ in Table 4.7.

TABLE 4.7
The Time-Domain Comprehensive Performance of Six Controllers

	$ITAE_{sum} = ITAE_{st} + ITAE_{ld}$		
Process	**Lag-Dominated**	**Lag-Delay Balanced**	**Delay-Dominated**
Proposed 1	3.189	41.76	61.01
Chen et al.	3.277	–	–
Chevalier et al.	3.815	49.36	–
Yumuk et al.	5.2544	46.13	85.83
Valrio et al.	51.77	74.6	70.15
Proposed 2	1.293	23.23	–

4.6 CONCLUSION

This chapter proposes an analytical design method for the FOPI$^{\lambda}$D$^{\mu}$ controller based on five frequency-domain specifications. Gain margin and phase margin guarantee the relative stability of control system. Gain crossover frequency influences the rising time of the step response. Flat phase keeps system response with the iso-damping property. Besides, through the frequency- and time-domain analysis, the impact of phase crossover frequency is fully exposed firstly. The design synthesis is given and summarized with illustration of an example. The designed control systems can accurately meet the user-specified frequency-domain specifications. Furthermore, the complete achievable region of four specifications is collected for all the FOPNTD systems. With the analytical design method and achievable region as the prior information, the implementation of the FOPI$^{\lambda}$D$^{\mu}$ controller in industry may increase widely. The pseudo codes of the designing scheme and collecting the achievable region for the FOPI$^{\lambda}$D$^{\mu}$ controller are summarized and presented. Note that codes of the design synthesis for the controller and collecting the achievable region are available for all the FOPID-type (including IOPID-type) controllers. Three simulation illustrations for lag-dominated, lag-delay balanced, and delay-dominated processes demonstrated that the designed FOPI$^{\lambda}$D$^{\mu}$ controller is applicable to a wider range of processes and achieves better step-tracking and anti-load disturbance performance over other controllers.

5 Fractional Order $PI^{\lambda}D^{\mu}$ Control Design with Bode's Ideal Cutoff Filter

5.1 INTRODUCTION

In Chapter 4, an analytical synthesis for the $FOPI^{\lambda}D^{\mu}$ controller with five parameters tuning based on five frequency-domain specifications is proposed for the first order plus normalized time delay (FOPNTD) systems. In order to expand upon this systematic approach, further research was conducted specifically targeting the speed servo system of permanent magnet synchronous motors. This chapter proposed a fractional order PID controller with a Bode's ideal cutoff (BICO) filter [72], which is named FOPID-BICO. The frequency-domain response of a BICO filter is compared with that of traditional first-order low-pass filter to show the advantages in magnitude and phase. A systematic analytical design method for the proposed FOPID-BICO controller is presented. The resulting speed servo system satisfies five frequency-domain specifications simultaneously: gain margin, phase margin, gain crossover frequency, phase crossover frequency, and a "flat phase" feature. Gain and phase margins ensure that the control system has the desired relative stability. Gain crossover frequency reflects the control bandwidth and response speed of the closed-loop system. The flat phase feature can guarantee the system is robust to loop gain variations. Moreover, by adjusting the phase crossover frequency in the design process, the frequency characteristics in both low frequency and high frequency can be adjusted for performance improvement simultaneously.

This chapter gives an interesting insight between the time-domain responses and the frequency-domain specifications of closed-loop servo systems. The main contributions of this chapter can be summarized as follows: (1) A FOPID-BICO controller is proposed for the permanent magnet synchronous motor (PMSM) speed servo system. (2) A systematic analytical design approach for the FOPID-BICO controller is firstly proposed to guarantee a certain relative stability margin, achieve satisfactory performances in rejecting load disturbance and noise, and realize robustness to system loop gain variations. (3) Experimental results on the PMSM speed servo system demonstrate that the proposed FOPID-BICO controller outperforms the optimal internal model control integer order PID (IMC-IOPID) controller, Bode's ideal transfer function fractional order PID (BITF-FOPID) controller, and the particle swarm optimization fractional order PID (PSO-FOPID) controller.

DOI: 10.1201/9781003397632-7

5.2 FOPI$^{\lambda}$D$^{\mu}$-BICO CONTROLLER DESIGN

5.2.1 Proposed Control Strategy

In this chapter, a PMSM speed controller is proposed, which is expressed as

$$C(s) = C_{FOPID}(s)C_{BICO}(s), \tag{5.1}$$

where $C_{FOPID}(s)$ is a FOPID controller, and the transfer function is given as (4.1). $C_{BICO}(s)$ is a BICO filter, which is introduced by Bode [72]. BICO can produce an optimal trade-off between the desired phase margin and the most rapid cut-off. The transfer function of the BICO is [72]

$$C_{BICO}(s) = \frac{1}{(\sqrt{1+(\frac{s}{\omega_{cf}})^2}+\frac{s}{\omega_{cf}})^r}, \tag{5.2}$$

where ω_{cf} is the cutoff frequency and $r \in (0,2)$ is a fractional order.

To clearly demonstrate the advantages of the BICO filter, the comparison of frequency-domain characteristics with a normal first-order low-pass filter is shown in Figure 5.1. The parameters of the BICO filter are $\omega_{cf} = 300$ rad/s and $r = 1$. The transfer function of the normal pass filter is $C_{nf}(s) = \frac{1}{0.0067s+1}$. Obviously, the amplitude in the pass-band of BICO filter is flat and followed by a rapid roll-off, and the phase is flatter than that of the one of the normal low pass filters below the cutoff frequency.

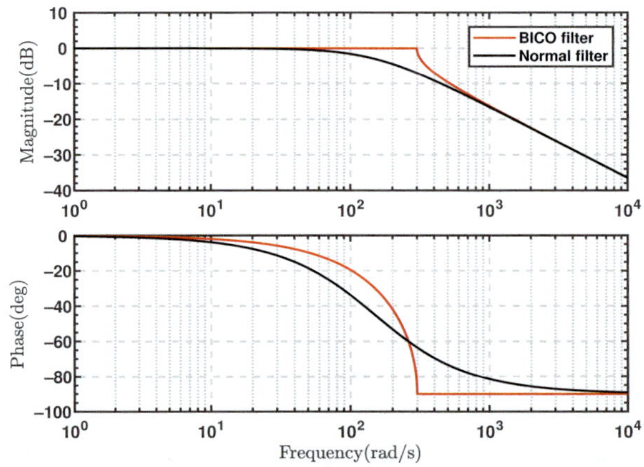

FIGURE 5.1 Bode plot comparison of a BICO filter and a normal low-pass filter.

5.2.2 An Analytical Design Method

To obtain the controller parameters satisfying the given frequency-domain specifications, the gain-phase margin tester (GPMT) is used and the control system is shown in Figure 5.2. The transfer function is given as (2.3).

The PMSM speed plant can be expressed as

$$G(s) = \frac{K}{s(Ts+1)}. \tag{5.3}$$

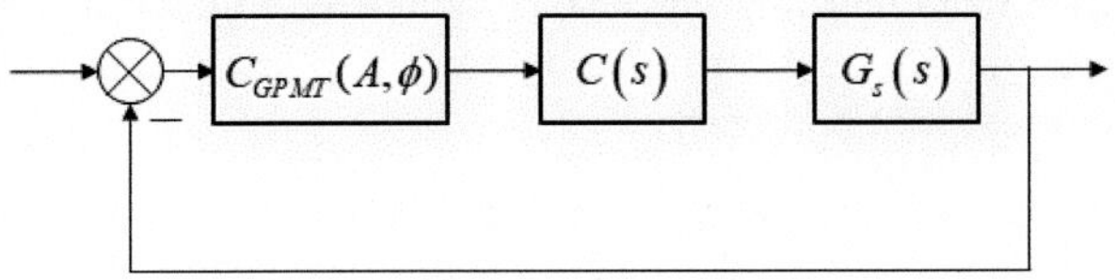

FIGURE 5.2 Control loop for parameter designing.

The closed-loop transfer function can be calculated as

$$G_c(s) = \frac{C(s)C_{GPMT}(A,\phi)G_s(s)}{1+C(s)C_{GPMT}(A,\phi)G_s(s)}. \tag{5.4}$$

Substituting (2.3), (5.1), (5.2), (5.3), into (5.4), the characteristic equation of (5.4) can be obtained:

$$D_c(s) = s^{1+\lambda}(Ts+1)(\sqrt{(\omega_{cf}^2+s^2)}+s)^r + AK\omega_{cf}^r(K_p s^{\lambda} + K_i + K_d s^{\lambda+\mu})e^{-j\phi}. \tag{5.5}$$

Letting $D_c(s = j\omega) = 0$, a line K_p-K_i can be obtained as ω increases from 0, which is named the CRB. Putting $s = j\omega$ into (5.5) yields

$$\begin{aligned} D_c(s=j\omega) &= (j\omega)^{(1+\lambda)}(jT\omega+1)(\sqrt{(\omega_{cf}^2+(j\omega)^2)}+j\omega)^r \\ &\quad + AK\omega_{cf}^r(K_p(j\omega)^{\lambda}+K_i+K_d(j\omega)^{\lambda+\mu})e^{-j\phi} = 0. \end{aligned} \tag{5.6}$$

Then,

$$\begin{aligned} B_5 + AK\omega_{cf}^r(C_4E+S_4F) = 0, \\ B_6 + AK\omega_{cf}^r(C_4F-S_4E) = 0, \end{aligned} \tag{5.7}$$

where

$$\begin{aligned}
&B_1 = T\omega^{\lambda+2}C_1 + \omega^{\lambda+1}C_2, B_2 = T\omega^{\lambda+2}S_1 + \omega^{\lambda+1}S_2, \\
&B_3 = \omega_{cf}^r C_3, B_4 = \omega_{cf}^r S_3, B_5 = B_1B_3 - B_2B_4, B_6 = B_2B_3 + B_1B_4, \\
&E = K_p\omega^{\lambda}C_5 + K_i + K_d\omega^{\lambda+\mu}C_6, F = K_p\omega^{\lambda}S_5 + K_d\omega^{\lambda+\mu}S_6, \\
&S_1 = \sin\frac{(\lambda+2)\pi}{2}, S_2 = \sin\frac{(\lambda+1)\pi}{2}, C_1 = \cos\frac{(\lambda+2)\pi}{2}, C_2 = \cos\frac{(\lambda+1)\pi}{2}, \\
&S_3 = \sin(r\theta), S_4 = \sin\phi, C_3 = \cos(r\theta), C_4 = \cos\phi, \\
&S_5 = \sin\frac{\lambda\pi}{2}, S_6 = \sin\frac{(\lambda+\mu)\pi}{2}, C_5 = \cos\frac{\lambda\pi}{2}, C_6 = \cos\frac{(\lambda+\mu)\pi}{2}, \theta = \operatorname{atan}\frac{\omega}{\omega_{cf}^2-\omega^2}.
\end{aligned}$$

Then, K_p and K_i can be solved through (5.7):

$$\begin{aligned}
K_p &= \frac{-(B_5S_4+B6C_4)}{AK\omega_{cf}^r\omega^{\lambda}S_5} - \frac{K_d\omega^{\mu}S_6}{S_5}, \\
K_i &= \frac{B_6 - B_5C_4S_4 - B_6C_4^2}{AKS_4\omega_{cf}^r} + \frac{B_5S_4C_5 + B_6C_4C_5}{AK\omega_{cf}^rS_5} + \frac{K_d\omega^{\lambda+\mu}S_6C_5}{S_5} - K_d\omega^{\lambda+\mu}C_6.
\end{aligned} \tag{5.8}$$

Given a fixed λ, μ, and K_d, the line of K_p-K_i can be plotted with (5.8).

Letting $D_c(s=0)=0$, a line K_p-K_i can also be obtained, which is named as the RRB. Next, the stable region can be determined by random point testing with the complex root boundary and real root boundary.

Remark 4. *Note that the $C_{GPMT}(A,\phi)$ can be used to obtain the controller parameters for a given gain margin or phase margin. In this chapter, to obtain the FOPID controller parameters for a given phase margin, one needs to set $A=1$.*

The design procedure can be summarized as:

1. Consider the frequency-domain specification phase margin ϕ_m: Fix $\phi=\phi_m$, $A=1$, and given any λ, μ, K_d, one can get a line of K_p-K_i as ω increases from 0 according to (5.8). All the points on the line of K_p-K_i satisfy the phase margin ϕ_m.

2. Consider the frequency-domain specification gain crossover frequency ω_{gc}: Given the $\omega=\omega_{gc}$, a point (K_p, K_i) can be determined on the above line through (5.8). Letting λ change from 0 to 2, one can get a curve satisfying two frequency-domain specifications phase margin, and gain crossover, which is named the ω_{gc}-ϕ_m curve.

3. Consider the frequency-domain specification "flat phase": The "flat phase" can guarantee the control system is robust to the loop gain variations. That means the phase margin has no change, which means the partial derivative of the phase with respect to ω is equal to 0. From (5.7), one can obtain the following equation:

$$\begin{aligned} S_4 &= \frac{B_6E-B_5F}{AK\omega_{cf}^r(E^2+F^2)}, \\ C_4 &= \frac{-B_5E-B_6F}{AK\omega_{cf}^r(E^2+F^2)}. \end{aligned} \tag{5.9}$$

Then,

$$\phi = \frac{B_5F-B_6E}{B_5E+B_6F}, \tag{5.10}$$

yields

$$\frac{\mathrm{d}\phi}{\mathrm{d}\omega} = \frac{(B_5^2+B_6^2)(EF'-E'F)+(B_5'B_6-B_5B_6')(E^2+F^2)}{(B_5E+B_6F)^2+(B_5F-B_6E)^2} = 0. \tag{5.11}$$

where

$$\begin{aligned}
&E' = \lambda K_p\omega^{\lambda-1}C_5+(\lambda+\mu)K_d\omega^{\lambda+\mu-1}C_6, \\
&F' = \lambda K_p\omega^{\lambda-1}S_5+(\lambda+\mu)K_d\omega^{\lambda+\mu-1}S_6, \\
&B_5' = B_1'B_3+B_1B_3'-B_2'B_4+B_2B_4', \quad B_6' = B_2'B_3+B_2B_3'+B_1'B_4-B_1B_4', \\
&B_1' = (\lambda+2)T\omega^{\lambda+1}C_1+(\lambda+1)\omega^{\lambda}C_2, \\
&B_2' = (\lambda+2)T\omega^{\lambda+1}S_1+(\lambda+1)\omega^{\lambda}S_2, \\
&B_3' = \omega_{cf}^r(-r\sin(r\mathrm{atan}\frac{\omega}{\sqrt{\omega_{cf}^2-\omega^2}}))(\omega_{cf}^2-\omega^2)^{(-\frac{1}{2})}, \\
&B_4' = \omega_{cf}^r(r\cos(r\mathrm{atan}\frac{\omega}{\sqrt{\omega_{cf}^2-\omega^2}}))(\omega_{cf}^2-\omega^2)^{(-\frac{1}{2})}.
\end{aligned}$$

Testing all the points on the above ω_{gc}-ϕ_m curve by (5.11), obtain the solution which is a certain point with the FOPID-BICO controller parameters. Sweeping all the μ in (0, 2), a curve named the flat phase curve can be obtained, and all the points on this curve satisfy the given phase margin, gain crossover frequency, and "flat phase."

4. Consider the frequency-domain specification phase crossover frequency ω_{pc}: The phases of $C_{FOPID}(s)$, $C_{BICO}(s)$ and $G_s(s)$ are

$$\mathrm{Arg}[C_{FOPID}(j\omega)] = \tan^{-1}\frac{D_1}{D_2}, \tag{5.12}$$

where

$$D_1 = -K_i\omega^{-\lambda}\sin\frac{\lambda\pi}{2} + K_d\omega^{\mu}\sin\frac{\mu\pi}{2},$$

$$D_2 = K_p + K_i\omega^{-\lambda}\cos\frac{\lambda\pi}{2} + K_d\omega^{\mu}\cos\frac{\mu\pi}{2}.$$

$$\mathrm{Arg}[G_s(j\omega)] = -\tan^{-1}(T\omega) - \frac{\pi}{2}. \tag{5.13}$$

$$\mathrm{Arg}[C_{BICO}(j\omega)] = \begin{cases} -r\mathrm{atan}\frac{\omega}{\sqrt{-\omega^2+\omega_{cf}^2}} & \omega < \omega_{cf} \\ \left(\sqrt{\left(\frac{\omega}{\omega_{cf}}\right)^2 - 1} + \frac{\omega}{\omega_{cf}}\right)^{-r} & \omega \geq \omega_{cf}, \end{cases} \tag{5.14}$$

The phase of the open-loop function $G_o(s)$ is

$$\mathrm{Arg}[G_o(j\omega_{pc})] = \mathrm{Arg}[C(j\omega_{pc})] + \mathrm{Arg}[G_s(j\omega_{pc})] + \mathrm{Arg}[C_{BICO}(j\omega_{pc})] = -\pi. \tag{5.15}$$

Testing all the points on the above flat phase curve by (5.15) to obtain the solution, this point satisfies four frequency-domain specifications: phase margin, gain crossover frequency, flat phase, and phase crossover frequency. Next, sweeping K_d in the reasonable range, a curve named ω_{gc}-ϕ_m-FP-ω_{pc} can be got. All the points on this curve meet phase margin, gain crossover frequency, flat phase, and phase crossover frequency.

5) Consider the frequency-domain specification gain margin A_m: The gains of $C_{FOPID}(s)$, $C_{BICO}(s)$, and $G_s(s)$ are

$$|C_{FOPID}(j\omega)| = \sqrt{D_1^2 + D_2^2}. \tag{5.16}$$

$$|G_s(j\omega)| = \frac{K}{\omega\sqrt{(T\omega)^2+1}}. \tag{5.17}$$

$$|G_{\mathrm{BICO}}(j\omega)| = \begin{cases} 1, & \omega < \omega_{cf} \\ \left(\sqrt{\left(\frac{\omega}{\omega_{cf}}\right)^2 - 1} + \frac{\omega}{\omega_{cf}}\right)^{-r}, & \omega \geq \omega_{cf} \end{cases} \tag{5.18}$$

The gain margin of the open-loop function $G_o(s)$ is

$$|G_o(j\omega_{pc})|_{dB} = -\log_{10}(|C_{FOPID}(j\omega_{pc})||G_s(j\omega_{pc})||G_{\mathrm{BICO}}(j\omega_{pc})|) = A_m. \tag{5.19}$$

Testing all the points on the above ω_{gc}-ϕ_m-FP-ω_{pc} curve by (5.19), the solution corresponding the controller parameters is the designed FOPID-BICO controller, which satisfies all the five given frequency-domain specifications.

5.3 CONTROLLER DESIGN EXAMPLE

To confirm the effectiveness of the proposed FOPID-BICO controller, the performance of the PMSM speed servo system using the proposed FOPID-BICO controller is compared with that using the optimized IMC-IOPID controller [73], BITF-FOPID controller [47], and PSO-FOPID controller [74].

In this section, the FOPID-BICO controller and other controllers used for comparison are designed for the PMSM speed plant (5.3) with $K = 14.7061$, $T = 0.0014$.

The IMC-IOPID controller is designed according to the proposed method in [73],

$$C_{IMC-IOPID}(s) = \frac{0.068(0.0014s+1)}{5.8564e-04s+0.0484}. \tag{5.20}$$

Given the same designed indicators $\omega_{gc} = 20$ rad/s and $\phi_m = 65°$, a BITF-FOPID controller is designed [47],

$$C_{BITF-FOPID}(s) = \frac{1}{s^{0.2778}}\frac{(1+0.0014s)}{0.3199}. \tag{5.21}$$

Next, a FOPID controller is designed using the PSO algorithm [74],

$$C_{PSO-FOPID}(s) = 1.2628 + \frac{9.8382}{s^{0.72458}} + 0.044165s^{0.52483}. \tag{5.22}$$

Then, two FOPID-BICO controllers are designed with the above proposed design procedure. One FOPID-BICO controller is designed with the same gain crossover frequency and phase margin:

$$C_{FOPID-BICO_1}(s) = (1.9353 + \frac{10.8935}{s^{1.325}} - 0.25s^{0.3075})(\sqrt{1+(\frac{s}{300})^2} + \frac{s}{300})^{-0.9}, \tag{5.23}$$

and the other is designed with bigger phase margin $\phi_m = 70^\circ$ and bigger gain crossover frequency $\omega_{gc} = 40$ rad/s:

$$C_{FOPID-BICO_2}(s) = (2.67 + \frac{18.5676}{s^{0.99}} + 0.001s^{0.7265})(\sqrt{1+(\frac{s}{300})^2} + \frac{s}{300})^{-0.9}. \tag{5.24}$$

To clearly illustrate the design process, the procedures with illustration for FOPID-BICO controller design are summarized in this section.

1. Considering the speed control plant, given the frequency-domain specifications $\phi_m = 70^\circ$, $\omega_{gc} = 40$ rad/s, $\omega_{pc} = 290$ rad/s, and $A_m = 18$ dB.
2. Fixed $\lambda = 0.7/0.9/1.1/1.3$, $\mu = 0.3$ and $K_d = 0.1$, the line of K_p-K_i can be plotted as shown in Figure 5.3 through (5.8) with $A = 1$ and $\phi = \phi_m$. All the points on this line meet $\phi_m = 70^\circ$.
3. Setting $\phi = \phi_m$ and $\omega_{gc} = 40$ rad/s in (5.8), a point can be fixed on the above line as the black dot shown in Figure 5.3.

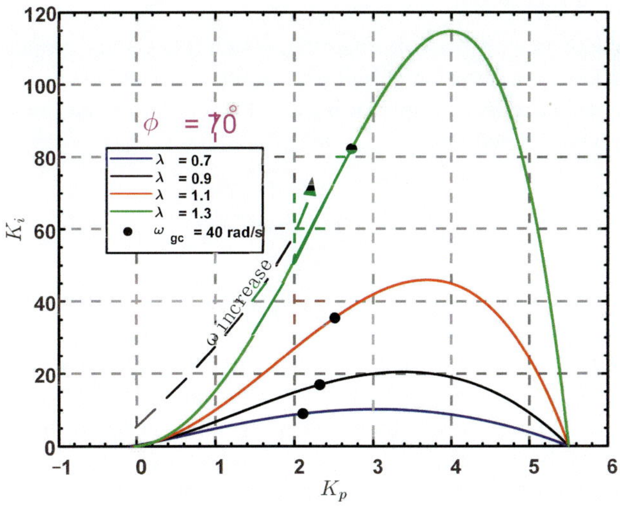

FIGURE 5.3 Phase margin $\phi_m = 70^\circ$ with $K_d = 0.1$, $\mu = 0.3$, and different λ.

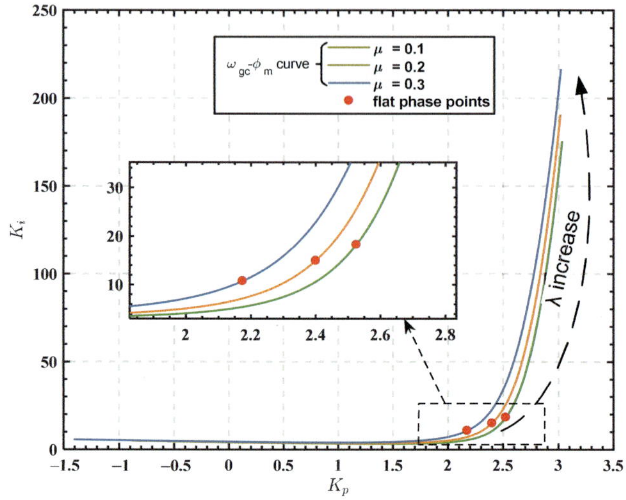

FIGURE 5.4 Phase margin $\phi_m = 70°$ and gain crossover frequency $\omega_{gc} = 40$ rad/s.

4. Sweeping λ in (0,2), a curve named ω_{gc}-ϕ_m can be plotted as shown in Figure 5.4. Then, one point can be determined as the red dot shown in Figure 5.4 by testing all the points using (5.11). This point is called the flat phase point.

5. Sweeping μ in (0, 2), the flat phase curve can be plotted as shown in Figure 5.5. Testing all the points on the flat phase curve by (5.15) with $\omega_{pc} = 290$ rad/s, a point can be fixed as square point shown in Figure 5.5. All the open-loop Bode plots associating with points at the flat phase curve are shown in Figure 5.6.

The complementary sensitivity function, which reflects performances of the speed-tracking and rejection of high-frequency noise and the load sensitivity function, which reflect the performance of rejection of the low-frequency load, can be presented as

$$T(s) = \frac{C(s)G_s(s)}{1+C(s)G_s(s)}, \tag{5.25}$$

$$E(s) = \frac{G_s(s)}{1+C(s)G_s(s)}. \tag{5.26}$$

The Bode plots of the functions $T(s)$ and $E(s)$ of all the flat phase points in Figure 5.5 are shown in Figures 5.7 and 5.8, respectively. It can be clearly seen from the Figure 5.5 that the resonance peak decreases as ω_{pc} increases, which means the overshoots decrease with ω_{pc} increasing. On the contrary, the ability to reject high-frequency noise increases as ω_{pc} decreases. From Figure 5.8, the ability to reject low-frequency load disturbance increases as ω_{pc} decreases.

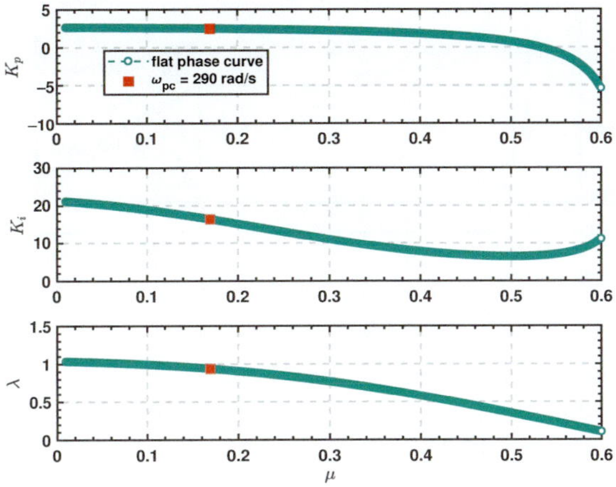

FIGURE 5.5 Flat phase curve.

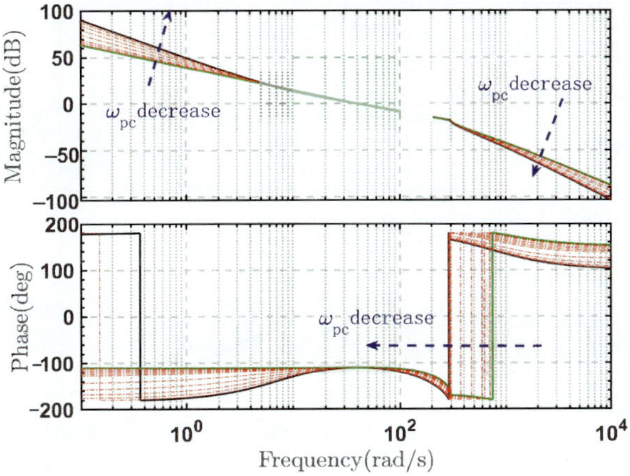

FIGURE 5.6 Bode plots of flat phase points.

6) Sweeping K_d in a $(-1, 1)$, ω_{gc}-ϕ_m-FP-ω_{pc} curve is presented in Figure 5.9. Testing all the points on the above curve by (5.19) with $A_m = 17$ dB, a point can be fixed as a blue dot. This point satisfied the given frequency-domain specifications $\phi_m = 60°$, $\omega_{gc} = 50$ rad/s, $\omega_{pc} = 290$ rad/s, and $A_m = 18$ dB. The open-loop Bode plot is shown in Figure 5.10.

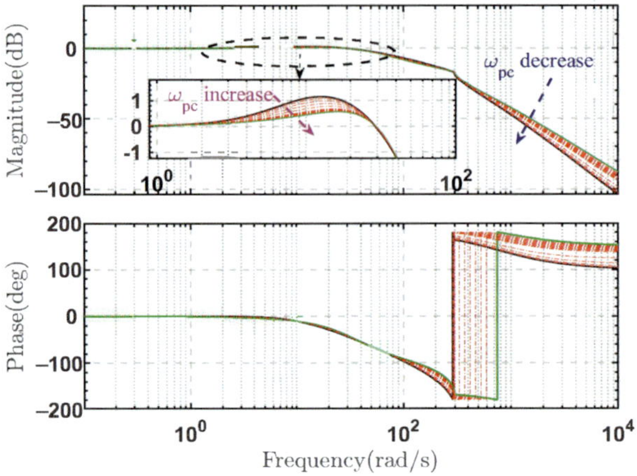

FIGURE 5.7 Complementary sensitivity Bode plots of flat phase points.

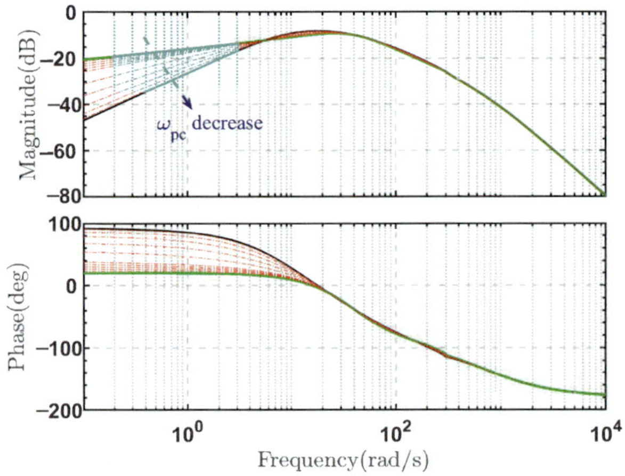

FIGURE 5.8 Load sensitivity Bode plots of flat phase points.

5.4 EXPERIMENTAL VALIDATION

5.4.1 Robustness to Loop Gain Variations

The gain variations of the PMSM speed plant are emulated by multiplying the controller's gain by 0.8 and 1.2, respectively. The speed responses experimental with loop gain variations using the designed $C_{FOPID-BICO_1}(s)$ are carried out, and the results are shown in Figure 5.11. From Figure 5.11, the overshoots of the designed FOPID-BICO control system with gain variations are close to each other, which demonstrates superior robustness to gain variations.

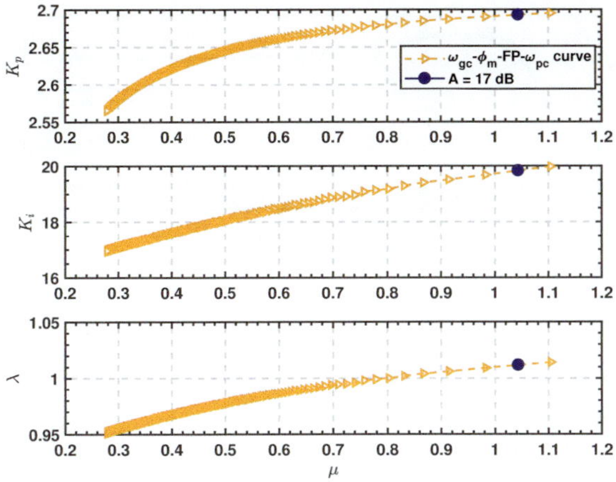

FIGURE 5.9 ω_{gc}-ϕ_m-FP-ω_{pc} curve.

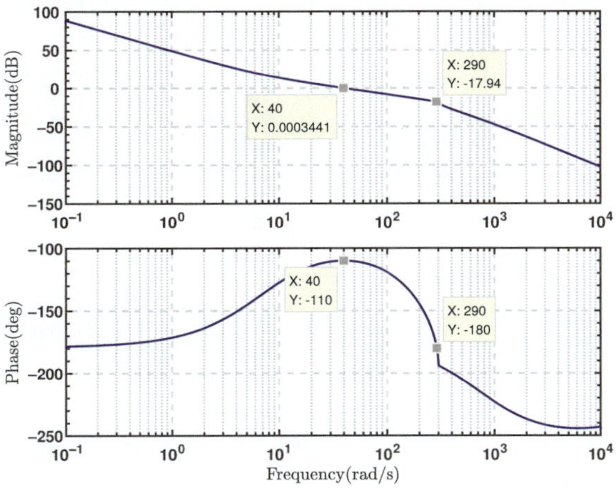

FIGURE 5.10 Open-loop Bode plot of the designed FOPID-BICO control system.

5.4.2 Comparison with Other Existing Controllers

Setting the reference speed as 1000 r/min, the step responses and load disturbance responses are carried out using the designed IMC-IOPID, BITF-FOPID, PSO-FOPID, and FOPID-BICO controllers. Figure 5.12a shows the experiment results of the motor actual speed, and Figure 5.12b shows the experiment results of the q-axis actual current. There is no overshoot, but there is steady state error (30 r/min) in the speed response of the IMC-IOPID control. After the external load is added to the system, the speed decreases about 40 r/min and cannot recover in the IMC-IOPID

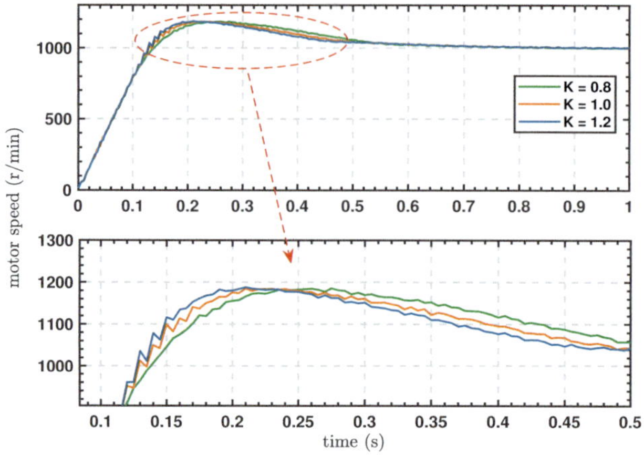

FIGURE 5.11 The speed responses of the FOPID-BICO$_1$ control system with gain variations (experiment).

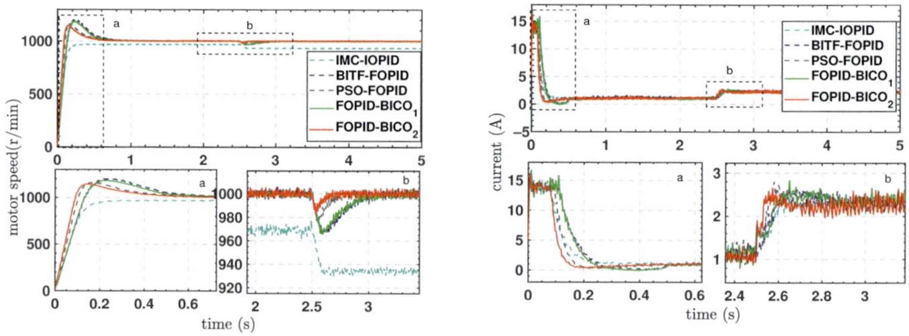

(a) Speed response (experiment). **(b)** Current response (experiment).

FIGURE 5.12 The speed and current responses.

control. The BITF-FOPID control system has a better load disturbance rejection ability than the IMC-IOPID control but has the biggest overshoot (20.2%). In the designed FOPID-BICO$_1$ control, the overshoot is smaller and the settling time is less than that of the BITF-FOPID control. In terms of load disturbance rejection, the designed FOPID-BICO$_1$ controller also achieves better performance than BITF-FOPID controller: The speed drop is 7.43% and the recovery time is 0.35 s. The PSO-FOPID control achieves better tracking and load disturbance rejection performance with less overshoot (16.9%) and speed drop (2.23%). In the designed FOPID-BICO$_2$ control system, the overshoot is the smallest (14.4 %) and the settling time is the shortest

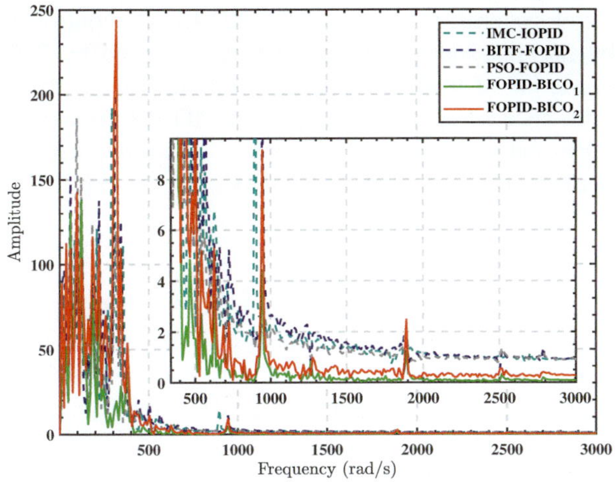

FIGURE 5.13 The FFT of speed-tracking error (experiment).

(0.486 s). When an external load is added, the speed drop of the FOPID-$BICO_2$ control system is 1.49%, and it takes 0.062 s to return to the reference speed. Figure 5.13 shows the FFT of speed-tracking error. The result indicates that the proposed controller can achieve smaller amplitude than other controllers. The time-domain characteristics of five control systems are summarized in Table 5.1.

TABLE 5.1
The Time-Domain Performance (Experiment)

	Step Tracking			Anti-Load Disturbance	
Controller	**Overshoot (%)**	**Rise Time (s)**	**Settling Time (s)**	**Speed Drop (%)**	**Recovery Time (s)**
IMC-IOPID	0 (steady state error)	0.22	0.22	3.59	cannot recovery
BITF-FOPID	20.2	0.128	0.675	3.44	0.375
PSO-FOPID	16.4	0.1	0.538	2.23	0.124
FOPID-$BICO_1$	17.9	0.128	0.655	3.41	0.33
FOPID-$BICO_2$	14.4	0.09	0.486	1.49	0.062

5.4.3 Discussion

The following can be concluded from the above experimental results: (1) The proposed FOPID-BICO controller and design method can guarantee that the system is robust to loop gain variations, which is a significant improvement to deal with the PMSM plant parameters uncertainty. (2) In terms of speed tracking, the designed

IMC-IOPID control system has steady-state error; compared with BITF-FOPID and PSO-FOPID control systems, the designed FOPID-BICO$_2$ control system reduces the overshoot by 28.71% and 12.2% and the settling time by 28% and 9.67%, respectively.(3) In terms of load disturbance rejection, compared with other three control methods, the speed drop of the designed FOPID-BICO$_2$ controller is reduced by 58.50%, 56.69%, and 33.18%. (4) The proposed controller achieves much better noise rejection performance than other controllers.

5.5 CHAPTER SUMMARY

This chapter proposed a FOPID-BICO controller with a BICO filter for a PMSM speed servo system. To overcome the difficulty of the parameters tuning for a FOPID controller, an analytical design method to satisfy five frequency-domain specifications is proposed. Gain margin and phase margin guarantee the system relative stability. Flat phase constraint provides superior robustness to system loop gain variation. Gain crossover frequency reflects the system control bandwidth representing the tracking response speed. Meanwhile, the effect of phase crossover frequency adjustment for loop-shaping including both low-frequency and high-frequency ranges is exposed with analysis and verification. The abilities of load disturbance rejection and noise rejection can be improved simultaneously by adjusting phase crossover frequency. The experimental results are carried out by comparing them with the optimized IMC-IOPID, BITF-FOPID, and PSO-FOPID controllers to verify that the proposed method can obtain a superior speed tracking performance and strong robustness to system loop gain variations and external load disturbances. Future works will be concentrated on the speed ripple due to the current measurement error, flux harmonics, and cogging torque under the proposed controller. Besides, the improvement of the bandwidth of the current and speed loops of the PMSM servo system is also a very promising research direction.

Part III

Fractional Order ADR Control

6 FO-ADRC with the Idea of a Cascaded FO Integrator

6.1 INTRODUCTION

From Chapter 2 to 5, the FOPID controllers can achieve better dynamic tracking performance. Robustness is also an important control performance in industrial applications. It was recognized at the stage of modern control that, regardless of the techniques, central to all control problems is the issue of uncertainties [13]. For example, Tsien denoted this as the problem of internal and external disturbances [75]; Horowitz discussed details about the uncontrollable external inputs to the plant and the partially ignored plant dynamics [76]. Many robust control approaches have been proposed in recent years. McFarlane [77] proposed a loop-shaping design method using H_∞ synthesis. An adaptive robust controller was proposed by Yin for an industrial robot [78]. The experimental study demonstrated that the adaptive robust controller can maintain a better trajectory tracking effect compared to a nonlinear controller. Other robust control methods include two-degrees-of-freedom control [79], disturbance-observer-based control [80] and so on [81]. ADRC [13], proposed by Jingqing Han, is the holistic solution to the above problems. ADRC treats the uncertainty of the plant dynamics as *internal disturbance* in addition to external disturbance. The internal and external disturbances make up *total disturbances* [82]. ESO, which can estimate the *total disturbances*, is the most important part of ADRC [82]. Many research results show that ADRC can achieve satisfied control performance [83–85]. However, the structure of ADRC is complex and limits ADRC applications in the industrial fields. Gao [14] proposed a linear active disturbance rejection control with two parameters: the observer and controller bandwidths, ω_o and ω_c, which extended the application of original ADRC from Han. Furthermore, Gao [86] proposed an ADRC involving the FO tracking differentiator, FO-PID controller, and FO-ESO for nonlinear FO systems. Li [87] proposed an integer order control scheme for FO systems using integer order ADRC. Simulation results demonstrated the effectiveness. A fractional order ADRC is proposed in [88] for FO systems, where the stability and frequency-domain characteristics are analyzed.

In this chapter, with the idea of cascaded FO integrator equivalence, an FO-ADRC design is proposed involving a fractional order ESO (FOESO) for a typical second-order motion platform. The FOESO converts the second-order plant of a PMSM speed system into a cascaded FO integrator platform ($\frac{1}{s^{2\alpha}}, 0 < \alpha < 1$). By using a simple proportional controller, a stable closed-loop feedback control system can be realized with enough phase margin and satisfying performance when the FO integrator order is properly selected based on the proposed design rule. This FO-ADRC

DOI: 10.1201/9781003397632-9

methodology is specially useful for real industrial applications with simple control design and implementation. An actual application of this FO-ADRC idea on the PMSM speed servo is demonstrated in this chapter. Simulation and experimental verifications are given to demonstrate the advantages of the proposed FO-ADRC over IO-ADRC and FOPI controller for the real industrial motion system. The proposed FO-ADRC achieves better robustness on high-frequency noise and controller gain variations than the IO-ADRC, and achieves better anti-load disturbance performance and robustness on plant gain variations than the FOPI controller.

6.2 FO-ADRC DESIGN

The transfer function of the speed servo system of the PMSM is

$$P(s)=\frac{Y(s)}{U(s)}=\frac{b}{s(s+a)}, \tag{6.1}$$

where $a=26.08$, $b=383.635$.

From equation (6.1), we can get

$$y^{(2\alpha)}=-\ddot{y}-a\dot{y}+y^{(2\alpha)}+\omega+bu=f_{fo}+b_0u, \tag{6.2}$$

where $b_0=b$, and

$$f_{fo}=-\ddot{y}-a\dot{y}+y^{(2\alpha)}+\omega. \tag{6.3}$$

Here, f_{fo} is the total disturbance including external disturbance ω, fractional dynamics $y^{(2\alpha)}$ where $0<\alpha<1$ and unknown internal dynamics $-\ddot{y}-a\dot{y}$.

Define $h=f_{fo}^{(\alpha)}$, then the state space form of (6.2) is

$$\begin{cases}\begin{bmatrix}x_1^{(\alpha)}\\x_2^{(\alpha)}\\x_3^{(\alpha)}\end{bmatrix}=\begin{bmatrix}0&1&0\\0&0&1\\0&0&0\end{bmatrix}\begin{bmatrix}x_1\\x_2\\x_3\end{bmatrix}+\begin{bmatrix}0\\b_0\\0\end{bmatrix}u+\begin{bmatrix}0\\0\\1\end{bmatrix}h,\\ y=\begin{bmatrix}1&0&0\end{bmatrix}\begin{bmatrix}x_1\\x_2\\x_3\end{bmatrix},\end{cases} \tag{6.4}$$

where x_1, x_2 are states, $x_3=f_{fo}$ is an extended state and f_{fo} can be estimated by FOESO as

$$\begin{cases}\begin{bmatrix}z_1^{(\alpha)}\\z_2^{(\alpha)}\\z_3^{(\alpha)}\end{bmatrix}=\begin{bmatrix}0&1&0\\0&0&1\\0&0&0\end{bmatrix}\begin{bmatrix}z_1\\z_2\\z_3\end{bmatrix}+\begin{bmatrix}0\\b_0\\0\end{bmatrix}u+L(y-\hat{y}),\\ \hat{y}=\begin{bmatrix}1&0&0\end{bmatrix}\begin{bmatrix}z_1\\z_2\\z_3\end{bmatrix},\end{cases} \tag{6.5}$$

where $L = [\beta_1 \quad \beta_2 \quad \beta_3]^{\mathrm{T}}$, and $\beta_1 = 3\omega_o$, $\beta_2 = 3\omega_o^2$, $\beta_3 = \omega_o^3$ according to [14], and z_1, z_2, and z_3 are the outputs of FOESO: z_1, z_2, and are the estimation of x_1, x_2, and x_3, respectively.

FOESO can estimate extended state f_{fo}, The designed control law is

$$u = \frac{-z_3 + u_0}{b_0}. \tag{6.6}$$

Bringing (6.6) into (6.2) yields a cascaded fractional order integrator form,

$$y^{(2\alpha)} \approx u_0. \tag{6.7}$$

A simple proportional controller $C_1(s) = k_{fp}$ can achieve stable control system because (6.7) is a stable plant. That is,

$$u_0 = k_{fp}(r - y). \tag{6.8}$$

Therefore, the desired closed-loop transfer function is,

$$\frac{Y(s)}{R(s)} = \frac{k_{fp}}{s^{2\alpha} + k_{fp}}. \tag{6.9}$$

6.3 FREQUENCY-DOMAIN ANALYSIS AND PARAMETER TUNING

The frequency-domain analysis and parameter tuning for the proposed FO-ADRC is presented in this section.

6.3.1 Frequency-Domain Analysis

From (6.5), through Laplace transform and solving the equations,

$$Z_1 = \frac{3\omega_o s^{2\alpha} + 3\omega_o^2 s^{\alpha} + \omega_o^3}{s^{3\alpha} + 3\omega_o s^{2\alpha} + 3\omega_o^2 s^{\alpha} + \omega_o^3} Y + \frac{b_0 s^{\alpha}}{s^{3\alpha} + 3\omega_o s^{2\alpha} + 3\omega_o^2 s^{\alpha} + \omega_o^3} U, \tag{6.10}$$

$$Z_2 = \frac{(3\omega_o^2 s^{\alpha} + \omega_o^3) s^{\alpha}}{s^{3\alpha} + 3\omega_o s^{2\alpha} + 3\omega_o^2 s^{\alpha} + \omega_o^3} Y + \frac{b_0 (s^{\alpha} + 3\omega_o) s^{\alpha}}{s^{3\alpha} + 3\omega_o s^{2\alpha} + 3\omega_o^2 s^{\alpha} + \omega_o^3} U, \tag{6.11}$$

$$Z_3 = \frac{\omega_o^3 s^{2\alpha}}{s^{3\alpha} + 3\omega_o s^{2\alpha} + 3\omega_o^2 s^{\alpha} + \omega_o^3} Y - \frac{b_0 \omega_o^3}{s^{3\alpha} + 3\omega_o s^{2\alpha} + 3\omega_o^2 s^{\alpha} + \omega_o^3} U, \tag{6.12}$$

where α is the order of FOESO. Choose fractional order $\alpha = 0.75$ to ensure that the phase margin of the control system is about 40°, choosing fractional order $\alpha = 0.75$. From Eqs. (6.1), (6.6), and (6.12) and Laplace transform, the transfer function of from u_o to y can be calculated as

$$P_{fo}(s) = \frac{Y(s)}{U_0(s)} = \frac{(s^{3\alpha} + \beta_1 s^{2\alpha} + \beta_2 s^{\alpha} + \beta) P(s)}{b_0 s^{3\alpha} + (\beta_3 P + b_0 \beta_1) s^{2\alpha} + \beta_2 b_0 s^{\alpha}}. \tag{6.13}$$

Setting ω_o with different values: 100, 300, 500, 700, 900 rad/s, the Bode plots of P_{fo} can be obtained in Figure 6.1. From the Bode diagrams, we know P_{fo} approximates $\frac{1}{s^{2\alpha}}$ better with the increase of ω_o, which is a cascaded fractional order integrator.

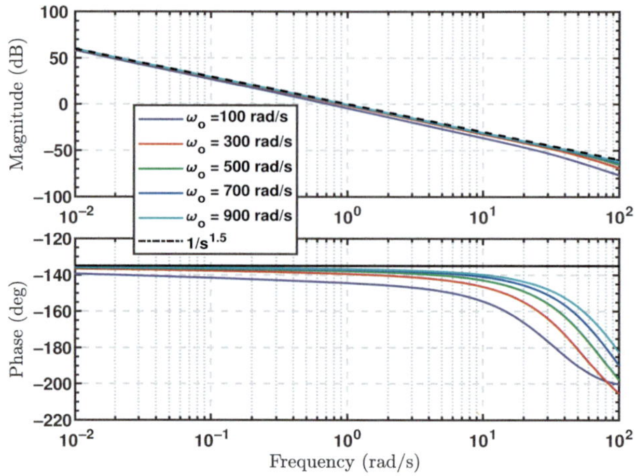

FIGURE 6.1 Bode plot of P_{fo}.

6.3.2 Parameters Tuning

For FO-ADRC, the open-loop transfer function G_{fo} is

$$G_{fo} = k_{fp}P_{fo}. \tag{6.14}$$

Setting the designed index as the gain crossover frequency $\omega_c = 10$ rad/s, the parameter k_{fp} can be tuned. Then, the phase margin ϕ_m can be calculated accordingly as in (6.15),

$$\begin{aligned} &\angle[G_{fo}(j\omega_c)] = \pi + \phi_m, \\ &|G_{fo}(j\omega_c)| = 1. \end{aligned} \tag{6.15}$$

For IO-ADRC, the open-loop transfer function G_{io} is

$$G_{io} = (k_{ip} + k_{id}s)P_{io}. \tag{6.16}$$

The parameters can be tuned applying two specifications: gain crossover frequency and phase margin,

$$\begin{aligned} &\angle[G_{io}(j\omega_c)] = -\pi + \phi_m, \\ &|G_{io}(j\omega_c)| = 1. \end{aligned} \tag{6.17}$$

Setting $\omega_o = 700$ rad/s, $\omega_c = 10$ rad/s, one can get $k_{fp} = 36.4469$, $\phi_m = 39°$, $k_{ip} = 86.5958$, and $k_{id} = 6.9713$. The open-loop Bode diagrams of FO-ADRC and IO-ADRC are shown in Figure 6.3a and two control systems have the same design specifications: $\omega_c = 10$ rad/s and $\phi_m = 39°$.

6.4 STABILITY ANALYSIS

First, we give a proof of the stability of FOESO; then, the stability analysis for entire closed-loop control system is presented.

6.4.1 Stability of FOESO

Let $e_i^{(\alpha)} = x_i^{(\alpha)} - z_i^{(\alpha)}, i = 1,2,3$. Combining (6.4) and (6.5), the error equation is

$$e_i^{(\alpha)} = A_e e_i + Eh, \tag{6.18}$$

i.e.

$$\begin{bmatrix} e_1^{(\alpha)} \\ e_2^{(\alpha)} \\ e_3^{(\alpha)} \end{bmatrix} = \begin{bmatrix} -\beta_1 & 1 & 0 \\ -\beta_2 & 0 & 1 \\ -\beta_3 & 0 & 0 \end{bmatrix} \begin{bmatrix} e_1 \\ e_2 \\ e_3 \end{bmatrix} + \begin{bmatrix} 0 \\ 0 \\ 1 \end{bmatrix} h. \tag{6.19}$$

The characteristic polynomial of A_e is

$$\lambda(s) = s^{3\alpha} + \beta_1 s^{2\alpha} + \beta_2 s^{\alpha} + \beta_3 = (s^{\alpha} + \omega_o)^3,$$

For the FO system above with FO α, two boundary lines for the stable region has the slope with $\pm\alpha\pi/2$. The stable region includes the left half-plane including an imaginary axis because of $0 < \alpha < 1$ [90]. Transforming the above FO system to w-plane by replacing $s^{\alpha} = w$, the transfer function is given by $\lambda(s) = (w + \omega_o)^3$. All the poles are determined at $-\omega_o$, i.e. at the left half-plane. Thus, the FOESO is bounded-input bounded-output (BIBO) stable.

6.4.2 Stability of the Closed-Loop Systems

Stability analysis for the FO-ADRC control system is studied in this subsection, and the conclusion is provided with the proof.

Lemma 6.1

An ordinary input/output relation can be written in a polynomial representation

$$\begin{aligned} P(\sigma)\xi &= Q(\sigma)u, \\ y &= R(\sigma)\xi, \end{aligned} \tag{6.20}$$

where $u \in \mathfrak{R}^{\tilde{m}}$ is the control, $\xi \in \mathfrak{R}^{\tilde{n}}$ is the partial state, and the $y \in \mathfrak{R}^{\tilde{p}}$ is the output; P, Q, and R are polynomial matrices in the variable σ of dimensions $\tilde{n} \times \tilde{n}$, $\tilde{n} \times \tilde{m}$, and $\tilde{p} \times \tilde{n}$, respectively; σ can be seen as the symbol of the usual derivative s^{α} when all initial conditions are zero.

For a proof of Lemma 1, see [91].

If the triplet (P, Q, R) of polynomial matrices is minimal, one has the following equivalence: The system is BIBO iff $det(P(\sigma)) \neq 0\ \forall \sigma, |arg(\sigma)| > \alpha\pi/2$. ■

Theorem 6.1

If choosing FO-ADRC parameters k_{fp}, α, and ω_o to assure

$$|arg(w_i)| > \lambda \frac{\pi}{2}, \tag{6.21}$$

where w_i are the ith root of the following equation:

$$\begin{aligned} &b\beta_1 w^{2pq+2p^2} + ab_0\beta_1 w^{pq+2p^2} + (b_0 + k_{fp})w^{3p^2} + (b\beta_3 + k_{fp}b\beta_1)w^{2p^2} \\ &+ (b_0\beta_2 + k_{fp}b\beta_2)w^{p^2} + k_{fp}b\beta_3 = 0, \end{aligned} \tag{6.22}$$

where $\lambda = 1/pq$, $p/q = \alpha$, p and q are positive integers, a and b are the plant parameters, k_{fp}, b_0, β_1, β_2, and β_3 are the FO-ADRC parameters. Then the FO-ADRC control system is bounded, as $t \to \infty$. ■

Proof. The closed-loop transfer function $G_c(s)$ of FO-ADRC control system is

$$G_c(s) = \frac{k_{fp}P_{fo}(s)}{1 + k_{fp}P_{fo}(s)}. \tag{6.23}$$

The characteristic equation of (6.23) is

$$\begin{aligned} N(s) &= b\beta_1 s^{2+2\alpha} + ab_0\beta_1 s^{1+2\alpha} + (b_0 + k_{fp}b)s^{3\alpha} + (b\beta_3 + k_{fp}b\beta_1)s^{2\alpha} \\ &+ (b_0\beta_2 + k_{fp}b\beta_2)s^{\alpha} + k_{fp}b\beta_3. \end{aligned} \tag{6.24}$$

Notate that

$$\alpha = \frac{p}{q},\ s^{\alpha} = s^{p/q}. \tag{6.25}$$

Thus,

$$\begin{aligned} N(s) &= b\beta_1 s^{2+2p/q} + ab_0\beta_1 s^{1+2p/q} + (b_0 + k_{fp}b)s^{3p/q} + (b\beta_3 + k_{fp}b\beta_1)s^{2p/q} \\ &+ (b_0\beta_2 + k_{fp}b\beta_2)s^{p/q} + k_{fp}b\beta_3. \end{aligned} \tag{6.26}$$

Notate $\lambda = 1/pq$, $w = s^{\lambda}$, then

$$\begin{aligned} N(w) &= b\beta_1 w^{2pq+2p^2} + ab_0\beta_1 w^{pq+2p^2} + (b_0 + k_{fp})w^{3p^2} + (b\beta_3 + k_{fp}b\beta_1)w^{2p^2} \\ &+ (b_0\beta_2 + k_{fp}b\beta_2)w^{p^2} + k_{fp}b\beta_3 = 0. \end{aligned} \tag{6.27}$$

According to Lemma 1, $\tilde{m} = \tilde{n} = 1$, since

$$|arg(w_i)| > \lambda \frac{\pi}{2}. \tag{6.28}$$

Thus, the closed-loop system is BIBO. □

6.5 SIMULATION ILLUSTRATION

The PMSM speed servo system can be controlled in MATLAB simulation, using the proposed FO-ADRC. By the performance comparisons with IO-ADRC, the advantages of the proposed FO-ADRC is illustrated clearly.

The parameters of the designed FO-ADRC are $k_{fp} = 36.4469$, $\alpha = 0.75$, and $\omega_o = 700$. Putting the parameters into (6.22), the pole plot can be found as shown in Figure 6.2, and the slopes of the two stable region boundary lines are $\pm\lambda\frac{\pi}{2}$ (corresponding to the $\pm\frac{\pi}{2}$ for the integer order systems). Thus, the system should be BIBO stable.

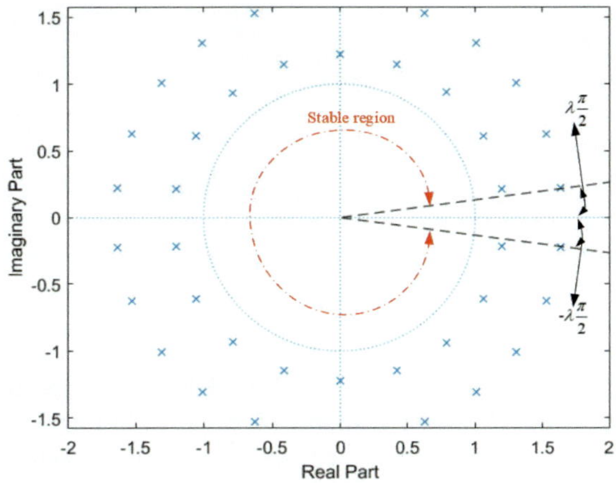

FIGURE 6.2 Pole plot of the designed FO-ADRC closed-loop control system.

PMSM speed response simulations are presented with reference speed 600 rpm, using the designed FO-ADRC and IO-ADRC, respectively. The simulation results are presented in Figure 6.3b. The speed response of the FO-ADRC system has smaller overshoot, oscillation, and shorter settling time. Thus, the FO-ADRC can achieve better step response performance over IO-ADRC.

As shown in Figure 6.3a, the phase of the FO-ADRC Bode plot near the gain crossover frequency point is flat, that is to say that the system is robust to loop gain variations, and the overshoots are basically equal with controller gain variations. Set the controller gain of FO-ADRC control system to be 80%, 100% and 120% of the nominal value. The step responses are presented in Figure 6.3c, and the FO-ADRC system has similar overshoots with different controller gains.

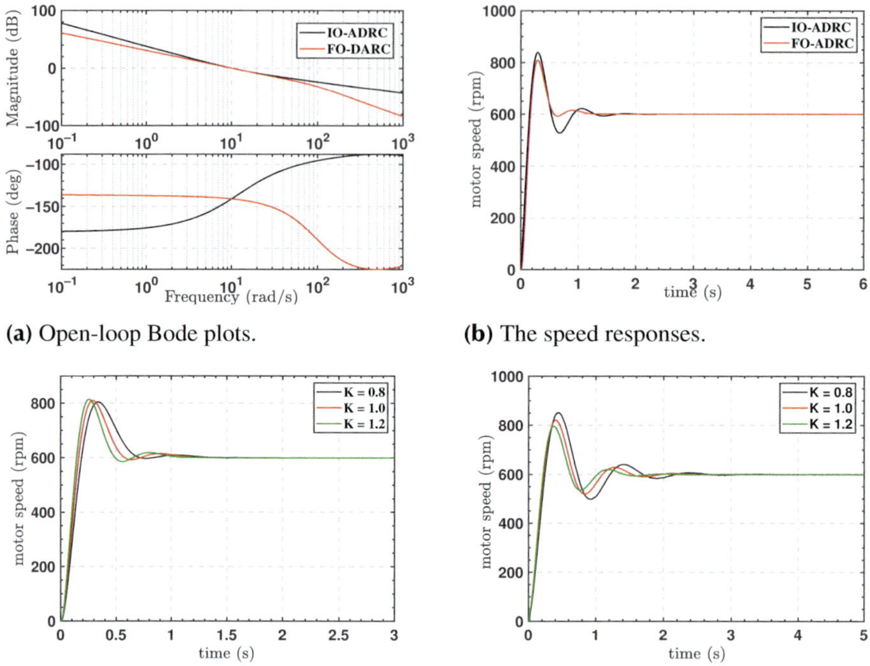

(**a**) Open-loop Bode plots.

(**b**) The speed responses.

(**c**) FO-ADRC system.

(**d**) IO-ADRC system.

FIGURE 6.3 Open-loop Bode plots, speed responses, and robustness comparison using the designed IO-ADRC and FO-ADRC.

For fair comparison, set the controller gain of IO-ADRC also to be 80%, 100%, and 120% of the nominal value. The step responses are shown in Figure 6.3d. As we can see from Figure 6.3d, the overshoots of the responses using IO-ADRC with different gains show obvious differences. Therefore, the FO-ADRC system achieves better robustness than the FO-ADRC system.

The proposed FO-ADRC does not contain a differential controller. Comparing the IO-ADRC with the PD controller, FO-ADRC does not amplify system noise and is much less sensitive to noise disturbance. The current responses and step responses are also compared in Figures 6.4 and 6.5, respectively, when system output contains noise. Obviously, FO-ADRC outperforms IO-ADRC.

6.6 EXPERIMENTAL VALIDATION

6.6.1 Step Response Comparison

The reference speed is set to 600 rpm, and PMSM speed is controlled, using the FO-ADRC and the IO-ADRC, respectively. The speed responses are presented in

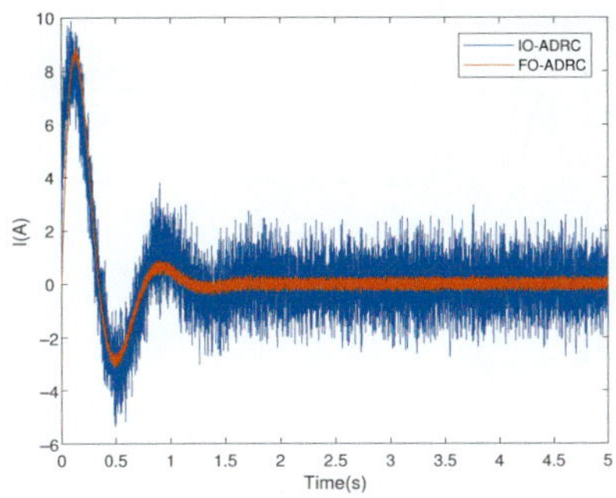

FIGURE 6.4 The current responses of IO-ADRC and FO-ADRC control systems (simulation).

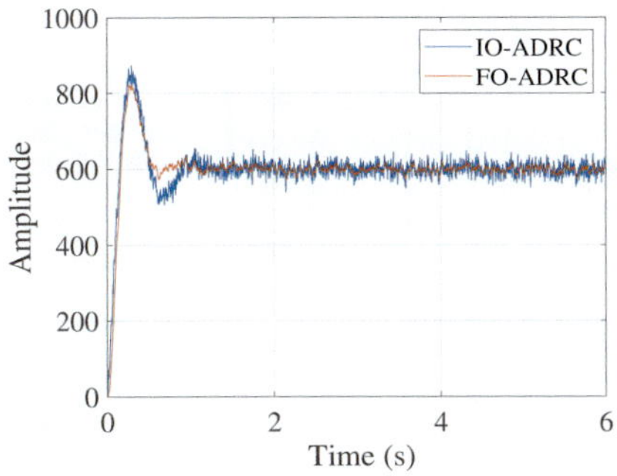

FIGURE 6.5 The speed responses of IO-ADRC and FO-ADRC control systems (simulation).

Figure 6.6, and the speed response of the FO-ADRC system has smaller overshoot, oscillation, and shorter settling time than the IO-ADRC system, which is similar to the simulation result. Therefore, the FO-ADRC system achieves better speed response performance.

6.6.2 Controller Gain Robustness Study

To verify the FO-ADRC is robust to controller gain variation, set the controller gain as 80%, 100%, and 120% of the nominal value. The motor speed response experiments are presented in Figure 6.7a,

The overshoots with different controller gains are similar in Figure 6.7a. Setting the same reference speed and controller gains, the speed response experiments of the IO-ADRC system are shown in Figure 6.7b. As we can see, it has poor robustness to controller gain compared to the FO-ADRC system.

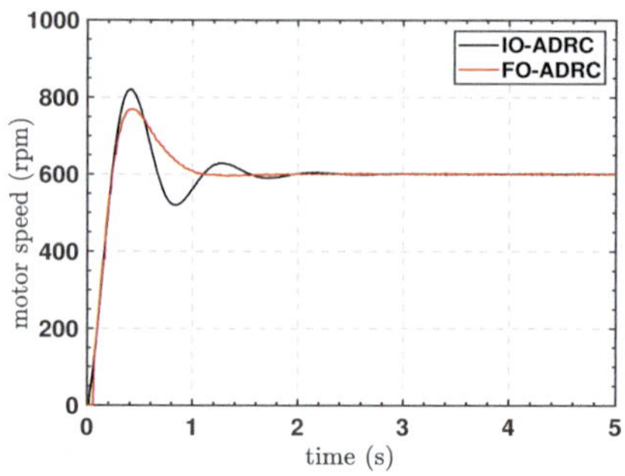

FIGURE 6.6 The speed responses of FO-ADRC and IO-ADRC (experiment).

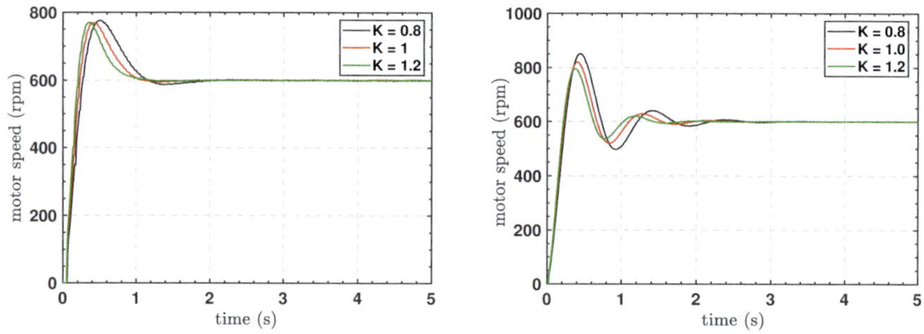

(a) FO-ADRC system. **(b)** IO-ADRC system.

FIGURE 6.7 Robustness comparison using the designed IO-ADRC and FO-ADRC.

6.7 CHAPTER SUMMARY

A FO-ADRC design methodology is proposed for a typical second-order motion control system based on a FOESO in this chapter. The FOESO converts the plant model into a cascaded FO integrator, which is a stable system. Therefore, a simple proportional controller design can achieve a stable closed-loop feedback control system. The feedback controller is designed according to the given gain crossover frequency, phase margin, and the bandwidth of FOESO. An actual application on the PMSM speed servo is studied by simulations and experiments. Simulation and experimental results verify that the proposed FO-ADRC achieves better speed response performance than IO-ADRC and FOPI controllers. Besides, the proposed FO-ADRC has outstanding robustness to controller and plant gain variations. Moreover, the proposed FO-ADRC is less sensitive to noise from the current experiment result compared with the experiment result of IO-ADRC.

7 Quasi-Fractional-Resonant-NESO Design for PMSM Servo Current Measurement Error

7.1 INTRODUCTION

Permanent magnet synchronous motors (PMSMs) have been widely applied in various industrial applications [92] [93]. However, harsh speed ripples caused by low-frequency torque ripple can deteriorate the whole performance of the PMSM drive system and further cause undesirable system vibrations at low speed. Thus, it is necessary and meaningful to improve the speed ripple suppression performance in industrial PMSM applications.

Without considering the impact of the load torque, the speed ripples mainly result from some non-ideal elements such as cogging torque [94], dead time of the inverter [95], current measurement error [96], and flux harmonics [94]. Among all the above reasons, current measurement error is the most critical one because the accuracy of real-time measurement for three-phase current cannot be satisfied using typical existing current sensors in high-precision PMSM servo system [97]. In the alternating current (AC) machine drive, the current measurement procedure usually includes polarity conversion, analog-to-digital (A/D) converter, sampling, low-pass filtering, and so on [97]. The zero voltage drift or offset of operational amplifiers and A/D converters in the electrical circuits cause current measurement offset error. The current scaling error is usually caused by the varying sampling resistance or operational amplifiers.

Many research studies have been conducted focusing on the current measurement error [97–104]. One is directly compensating the current measurement error. The scaling error can be withdrawn and compensated by the high-frequency signal injection [98]. However, the motor appears to feature high-frequency chatting, and the offset error still exists. In [99], the current measurement error is compensated by using the commanded voltage reference of the current controller, which leads to a slow error convergence. A compensation method was proposed to obtain the current measurement error from the d/q-axis and then subtract it from the measured current [97]. This method has an assumption that the derivatives of d/q-axis currents can be neglected, which is difficult to meet in real industrial processes. In [100], the current sensor in the DC bus is applied to calibrate phase current sensors, which

DOI: 10.1201/9781003397632-10

needs additional hardware. Thus, it increases cost and is less preferred than software improvement. Another type of current measurement error compensation method is based on the control strategy, such as resonant control [96], repetitive control [105], model predictive control (MPC) [106] and so on [103] [107]. In [96], a proportional-integral-resonant (PIR) controller in the speed control loop was proposed to deal with the speed ripple at low-speed due to the current measurement error and cogging effect. Studies in [106] proposed a cascade MPC control structure to suppress the resulting speed ripple by current measurement error.

Extended state observer (ESO) was introduced by Professor Han to estimate the total disturbance includes external disturbances and internal uncertainties [13]. The studies of normalized ESO (NESO) have verified the NESO-based controllers can achieve desired control performance both in step-tracking and disturbance rejection in a PMSM speed servo system [108]. However, for the torque ripple resulting from current measurement error, the performance is not satisfactory, especially at low speed. Recently, fractional calculation has attracted many scholars in the field of control engineering [109–112]. Various real physical systems have been proved to be described better by fractional differential equations, and fractional order (FO) controllers can obtain superior performance than the integer order ones [112–114]. Traditional resonant controllers can effectively deal with the disturbance with specific frequencies. With the development of the fractional order calculus, an ideal fractional-resonant (FR) controller was proposed, and it was verified that the ideal FR controller provides more robustness against fluctuations of center frequency and improves the time-domain transient performance than the traditional integer-resonant (IR) controller [115]. However, the proposed ideal FR controller is quite sensitive to the accuracy of resonant frequency, and its infinite gain may cause stability.

Aiming to solve the above problems, this chapter proposes a non-ideal FR controller which is named the quasi-fractional-resonant (QFR) controller and is more applicable in practice over the ideal FR. Meanwhile, a PMSM speed controller is proposed, including a proportional derivative (PD) controller, a QFR controller and a NESO. This controller can effectively deal with the low-speed ripple. A systematic design scheme which is based on frequency-domain analysis for the proposed controller is introduced.

7.2 ANALYSIS AND EFFECT OF CURRENT MEASUREMENT ERROR

The current measurement process mainly includes sampling, low-pass filtering, polarity conversion, and calculation. The measurement current can be expressed as

$$i_m = Ki + i_o, \tag{7.1}$$

where i_m is the measurement current, i is the actual current, K is the scaling gain, and i_o is the offset error.

According to (7.1), the current measurement error is mainly caused by two factors. One is the scaling error, which is caused by the nonlinearity of the current sensor itself, the matching circuit between the current sensor and A/D input, and the varying sampling resistance caused by temperature rise. Another is the offset error, which is caused by unbalanced positive and negative feeding voltages of the sensor and zero voltage drift of the analog devices.

A. Scaling Error

Considering scaling error, the measurement current is

$$\begin{aligned} i_{ma} &= K_a i_a, \\ i_{mb} &= K_b i_b, \end{aligned} \tag{7.2}$$

where K_a and K_b are the scaling gains of currents in phase a and phase b, respectively; i_{ma} and i_{mb} are the measurement current values, respectively; and i_a and i_b are the actual current of phase a and phase b, respectively.

The scaling current errors of the measured synchronous d-axis and q-axis currents can be derived as [94]

$$\begin{aligned} \triangle i_d &= \frac{\sqrt{3}}{3}(K_b - K_a)I\sin(2\theta_e + \alpha) + \frac{2 - K_a - K_b}{2}I, \\ \triangle i_q &= \frac{\sqrt{3}}{3}(K_b - K_a)I\sin(2\theta_e + \alpha) - \frac{\sqrt{3}}{6}(K_a - K_b)I, \end{aligned} \tag{7.3}$$

where $\triangle i_d = i_{md} - i_d$ and $\triangle i_q = i_{mq} - i_q$; i_d and i_q are actual current values of d-axis and q-axis, respectively; i_{md} and i_{mq} are the measurement values for i_d and i_q, respectively; θ_e is the rotor electrical angle; I is the amplitude of phase current; and α is determined by the angle between the stator current vector and the rotor position.

Eq. (7.3) shows that the motor steady-state speed exists two times the fundamental component due to the scaling error.

B. Offset Error

Considering the offset error, the measurement current is

$$\begin{aligned} i_{ma} &= i_a + i_{oa}, \\ i_{mb} &= i_b + i_{ob}, \end{aligned} \tag{7.4}$$

where i_{oa} and i_{ob} are the offset errors in phase a and phase b.

Transforming (7.4) into synchronous rotating coordinates yields

$$\begin{aligned} \triangle i_d &= \frac{2}{\sqrt{3}}\sqrt{i_{oa}^2 + i_{oa}i_{ob} + i_{ob}^2}\sin(\theta_e + \beta), \\ \triangle i_q &= \frac{2}{\sqrt{3}}\sqrt{i_{oa}^2 + i_{oa}i_{ob} + i_{ob}^2}\sin(\theta_e + \beta), \end{aligned} \tag{7.5}$$

where β is the angle between stator current vector and rotor position.

It can be seen from (7.5) that the current feedback channel will introduce disturbance the same as the fundamental frequency. Moving $\triangle i_q$ to the front of the current loop, the disturbance can be treated in the speed loop and solved by the speed controller.

7.3 CONTROLLER DESIGN

7.3.1 QFR-NESO Design

Figure 7.1 shows the PMSM control method based on the proposed PD-QFR-NESO speed controller, which includes a PD controller and a QFR-NESO. The structure of the QFR-NESO is shown in Figure 7.2, QFR can be expressed as

$$R_i(s) = \frac{2k_{ri}\omega_{ci}s^{\mu_i}}{s^{2\mu_i} + 2\omega_{ci}s^{\mu_i} - (j\omega_{ri})^{2\mu_i}}, i = 1,2, \tag{7.6}$$

where k_{ri} is the resonant coefficient, ω_{ri} is the resonant frequency, ω_{ci} is the cut-off frequency, and μ_i is the fractional order. The structure of the QFR is shown in Figure 7.3. If set $\mu_i = 1$, the R_i will be an integer order quasi-resonant (QR) controller [116].

A. QFR-NESO Design

Considering the disturbance, the plant (6.1) can be transferred as

$$\ddot{y} = -a\dot{y} + b_0 u + f, \tag{7.7}$$

where y is the output, u is the control signal, and $b_0 = b$. The total disturbance f is formulated by

$$f = f_i + f_l + f_c, \tag{7.8}$$

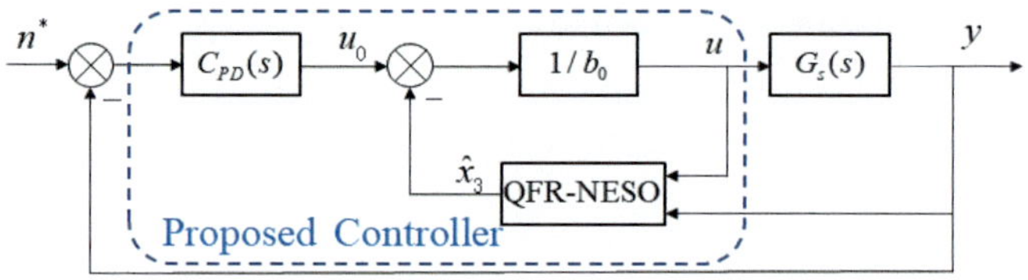

FIGURE 7.1 Block diagram of the proposed PMSM speed control system.

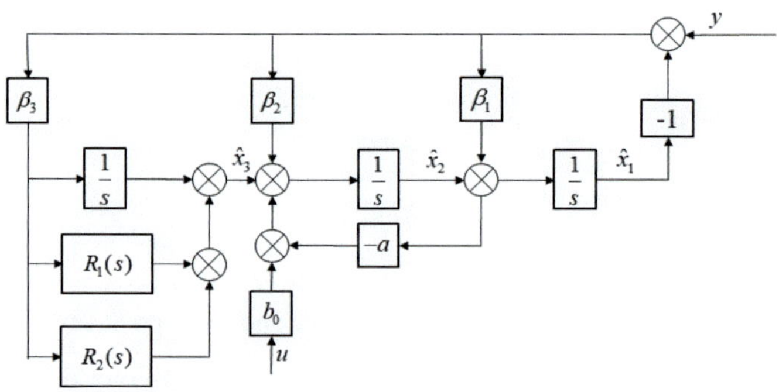

FIGURE 7.2 Block diagram of the proposed QFR-NESO.

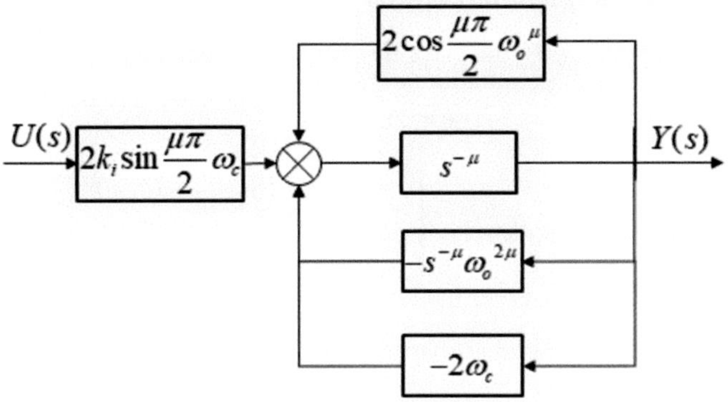

FIGURE 7.3 Block diagram of the proposed QFR.

where $f_i = (b - b_0)u$ represents the internal disturbance due to the mismatch between b and b_0, f_l is the load disturbance, f_c represents disturbances due to the current measurement error, and the frequencies of f_c are available according to the reference speed.

To estimate the total disturbance, f is extended to be a state variable which is hypothetically differentiable ($\dot{f} = h$). By defining $\boldsymbol{x} = [x_1, x_2, x_3] = [y, \dot{y}, f]$, the state space representation of (7.7) can be written as

$$\begin{aligned} \dot{\boldsymbol{x}} &= \boldsymbol{A}x + \boldsymbol{B}u + \boldsymbol{E}h \\ y &= \boldsymbol{C}\boldsymbol{x}, \end{aligned} \tag{7.9}$$

where $\boldsymbol{A} = \begin{bmatrix} 0 & 1 & 0 \\ 0 & -a & 1 \\ 0 & 0 & 0 \end{bmatrix}$, $\boldsymbol{B} = \begin{bmatrix} 0 \\ b_0 \\ 0 \end{bmatrix}$, $\boldsymbol{E} = \begin{bmatrix} 0 \\ 0 \\ 1 \end{bmatrix}$, $\boldsymbol{C} = \begin{bmatrix} 1 & 0 & 0 \end{bmatrix}$.

Based on (7.9), the QFR-NESO can be formulated as follows:

$$\dot{\hat{\boldsymbol{x}}} = \boldsymbol{A}\hat{\boldsymbol{x}} + \boldsymbol{B}u + \boldsymbol{L}(y - \hat{y}), \tag{7.10}$$

where $\hat{\boldsymbol{x}} = [\hat{x}_1, \hat{x}_2, \hat{x}_3]$ and $\hat{\boldsymbol{x}}$ will track $\boldsymbol{x}$, $\boldsymbol{L}$ is the observer gain vector and $\boldsymbol{L} = [\beta_1, \beta_2, \beta_3(1 + R(s)s)]$, and $R(s) = R_1(s) + R_2(s)$.

B. Control Law

The following control law is designed for disturbance compensation,

$$u = \frac{-\hat{x}_3 + u_0}{b_0}, \tag{7.11}$$

where u_0 is the output of the PD controller and the transfer function is

$$C_{PD}(s) = K_{sp} + K_{sd}s, \tag{7.12}$$

where K_{sp} and K_{sd} are the proportional and differential gains, respectively.

7.3.2 Disturbance Rejection Analysis

Subtracting (7.10) from (7.9), the error dynamic is obtained as

$$\dot{\boldsymbol{e}} = \boldsymbol{A}_e\boldsymbol{e} + \boldsymbol{E}h, \tag{7.13}$$

where $\boldsymbol{e} = [e_1 \ \ e_2 \ \ e_3]$, $\boldsymbol{A}_e = \begin{bmatrix} -\beta_1 & 1 & 0 \\ -\beta_2 & -a & 1 \\ -\beta_3(1+R(s)s) & 0 & 0 \end{bmatrix}$, $[\beta_1, \beta_2, \beta_3] = [3\omega_o, 3\omega_o^2, \omega_o^3]$, ω_o is the observer bandwidth.

Thus, the transfer function of disturbances estimation error can be obtained as

$$f_e(s) = \frac{e_3(s)}{f(s)} = \frac{s^3 + (\beta_1 + a)s^2 + (\beta_2 + a\beta_1)s}{s^3 + (\beta_1 + a)s^2 + (\beta_2 + a\beta_1)s + \beta_3(1 + R(s)s)}. \tag{7.14}$$

The Bode plots of $f_e(s)$ with different ω_o are shown in Figure 7.4. The other parameters of QFR-NESO are set as $\omega_{r1} = 26.18$, $\omega_{r2} = 52.36$, $\omega_{ci} = 0.01\omega_{oi}$, $\mu_1 = \mu_2 = 1.0$, and $k_{r1} = k_{r2} = 2$. The Bode plots indicate that the proposed QFR-NESO can track load disturbance and periodic disturbances, and the ability of disturbances rejection increases with the ω_o increasing. The Bode plots of $f_e(s)$ with different k_{ri}, ω_{ci}, and μ_i are shown in Figures 7.5, 7.6, and 7.7. It can be seen that k_{ri} affects the depth of specific frequency and ω_{ci} can widen the bandwidth. μ_i as the additional tuning knob has advantages and is more flexible over the integer order QR controller with only $\mu_i = 1$.

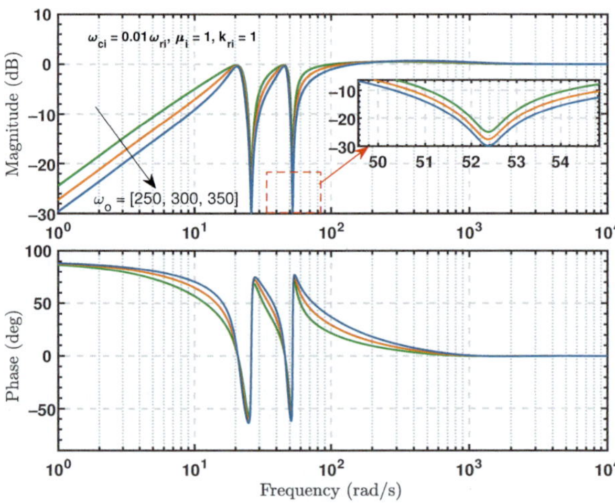

FIGURE 7.4 Bode plot of the transfer function $f_e(s)$ with different ω_o.

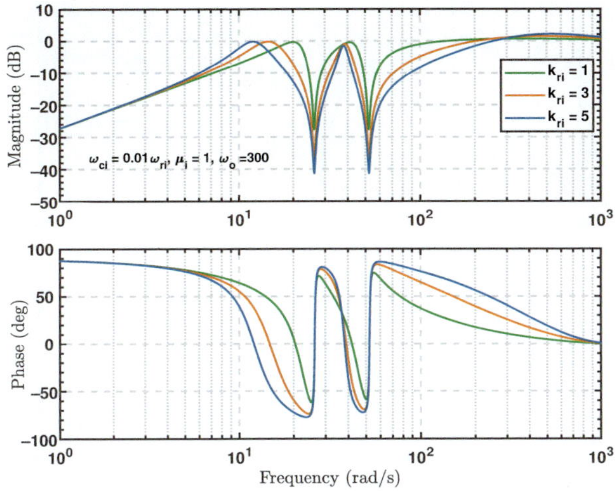

FIGURE 7.5 Bode plot of the transfer function $f_e(s)$ with different k_{ri}.

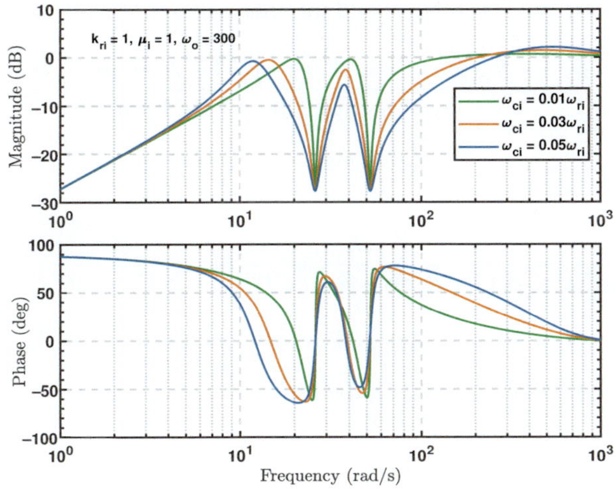

FIGURE 7.6 Bode plot of the transfer function $f_e(s)$ with different ω_{ci}.

7.3.3 Parameters Tuning

In this chapter, an output-error-based modeling method is adopted to obtain the PMSM parameters [117]. The identification results are $K_c = 1/19.2$, $K_v = 213.667$, $G_e(s) = \frac{1}{Rs+L} = \frac{1}{0.0024s+0.4583}$, $G_m(s) = \frac{2518.3}{s}$.

1. K_{sp} and K_{sd}

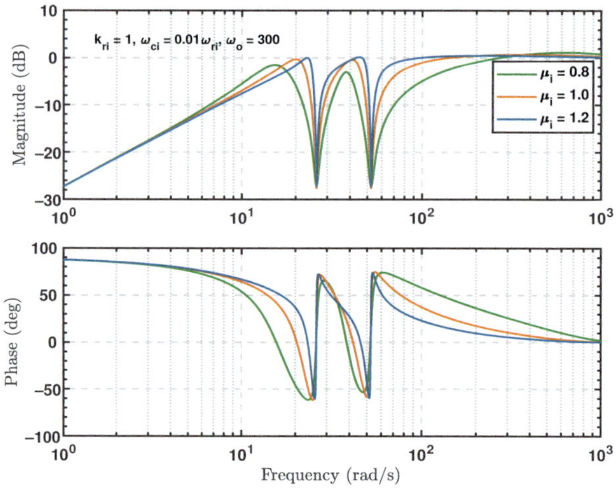

FIGURE 7.7 Bode plot of the transfer function $f_e(s)$ with different μ_i.

According to (7.10), $\hat{X}_3(s)$ can be solved through Laplace transform as

$$\hat{X}_3(s) = \frac{\beta_3(1+R(s)s)(s^2+as)Y(s) - \beta_3(1+R(s)s)b_0U(s)}{s^3+(a+\beta_1)s^2+(a\beta_1+\beta_2)s+\beta_3(1+R(s)s)}, \tag{7.15}$$

then, combining (7.15) and (7.11), the transfer function from u_0 to y can be expressed as

$$P(s) = \frac{G_s(s)F(s)}{b_0F(s)+\beta_3(1+R(s)s)(s^2+as)G_s(s) - b_0\beta_3(1+R(s)s)}, \tag{7.16}$$

where $F(s) = s^3+(a+\beta_1)s^2+(a\beta_1+\beta_2)s+\beta_3(1+R(s)s)$.

Substituting $P(s) = \frac{b}{s(s+a)}$ into (7.16), the result is

$$P(s) = \frac{1}{s(s+a)}. \tag{7.17}$$

Thus, the PMSM control system can be simplified as shown in Figure 7.8. The PD controller (7.12) can be designed following two frequency-domain specifications as follows:

i. Gain crossover frequency

$$|C_{PD}(j\omega)P(j\omega)||_{\omega=\omega_{sgc}} = 1, \tag{7.18}$$

where ω_{sgc} is the gain crossover frequency of the speed loop.

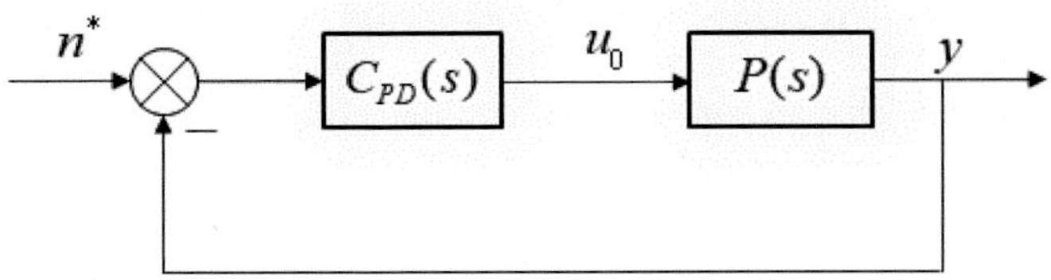

FIGURE 7.8 Feedback control system with a simplified plant.

ii. Phase margin

$$\angle(C_{PD}(j\omega)P(j\omega))|_{\omega=\omega_{sgc}} = \phi_m - \pi, \tag{7.19}$$

where ϕ_m is the desired phase margin.

The phase and gain of the plant (7.17) can be obtained as

$$\angle(P(j\omega)) = -\tan^{-1}(\frac{\omega}{a}) - \frac{\pi}{2}. \tag{7.20}$$

$$|P(j\omega)| = \frac{1}{\omega\sqrt{\omega^2 + a^2}}. \tag{7.21}$$

The phase and gain of the PD controller (7.12) are

$$\angle(C_{PD}(j\omega)) = \tan^{-1}\frac{K_{sd}\omega}{K_{sp}}, \tag{7.22}$$

$$|C_{PD}(j\omega)| = \sqrt{K_{sp}^2 + (K_{sd}\omega)^2}. \tag{7.23}$$

The gain margin of the open-loop function is

$$|C_{PD}(j\omega)P(j\omega)||_{\omega=\omega_{sgc}} = \frac{\sqrt{K_{sp}^2 + (K_{sd}\omega)^2}}{\omega\sqrt{\omega^2 + a^2}}. \tag{7.24}$$

The phase margin of the open-loop function is

$$\angle(C_{PD}(j\omega)P(j\omega))|_{\omega=\omega_{sgc}} = -\tan^{-1}(\frac{\omega}{a}) - \frac{\pi}{2} + \tan^{-1}\frac{K_{sd}\omega}{K_{sp}}. \tag{7.25}$$

Thus, the parameters K_{sp} and K_{sd} can be obtained according to the above two specifications. Given the gain crossover frequency $\omega_{sgc} = 60$ rad/s and phase margin $\phi_m = 80°$, the PD controller can be obtained as

$$C_{PD}(s) = 59713.6 - 114.56s. \tag{7.26}$$

2. ω_o, ω_{ci}, k_{ri}, ω_{ri} and μ.

ω_o reflects the ability of rejecting the load disturbance, but the noise will be amplified with ω_o increasing. Thus, ω_o can be determined according to the practical application requirements. In this chapter, ω_o is set as 300 rad/s. $\omega_{ci} = 1\% \cdot \omega_{ri}$ is set as usual, to deal with a typical $\pm 1\%$ variation of the fundamental frequency. ω_{ri} equals to the frequencies of the disturbances due to the current measurement error and yields

$$\omega_{r1} = \omega_e = p\omega_m = \frac{pn\pi}{30}\omega_{r2} = 2\omega_{r1}, \tag{7.27}$$

where $p = 5$ is the PMSM pole pairs. For example, $\omega_{r1} = 26.18$ rad/s and $\omega_{r2} = 52.36$ rad/s if $n = 50$ rpm.

To set the parameters of k_{ri}, some simple experimental tests need to be done. Because of the speed ripple problem is particularly prominent at low speeds, experimental tests with 50/60/70/80 rpm speed reference tracking are carried out. The steady-state speed responses and fast Fourier transform (FFT) are shown in Figures 7.10–7.13 which show that the total speed ripple due to scaling error and offset error is about 18 rpm and the peak disturbance is about 12 rpm. Requesting that the amplitude of the disturbance needs to be reduced to 0.1 rpm, the depth of the resonant frequency should be

$$20\log_{10}\frac{0.1 \text{ rpm}}{12 \text{ rpm}} = -41.58 \text{ dB}. \tag{7.28}$$

Thus, k_{ri} can be determined by plotting the Bode plot of $f_e(s)$ with different k_{ri} in MATLAB to satisfy -41.58 dB in the resonant frequency, and μ_i can be a flexible knob. According to the experimental results and Figure 7.7, μ_i is set as 0.8 in this chapter to deal with wider frequency disturbances.

7.4 EXPERIMENTAL VALIDATION

To confirm the effectiveness of the proposed PD-QFR-NESO controller, the performance of the PMSM speed servo system using the proposed controller is compared with that using the well-known optimal adaptive PIR controller [96]. The proposed controller in [96] includes a PI controller and an adaptive resonant controller to deal with current measurement error disturbances and is applied in the PMSM speed loop.

7.4.1 Tracking Performance

In this subsection, tracking performances of PMSM speed under the proposed controller and the adaptive PIR controller are thoroughly compared and studied. Setting the reference speed as 1000 rpm, the speed responses are shown in Figure 7.9a, using the proposed controller and adaptive PIR controller.

From Figure 7.9a, the proposed controller achieves no overshoot and the adaptive PIR controller has bigger overshoot (13.5%). In terms of the settling time, the proposed controller achieves 2.57 times faster than the adaptive PIR controller. The time-domain performances of two control systems are summarized in Table 7.1.

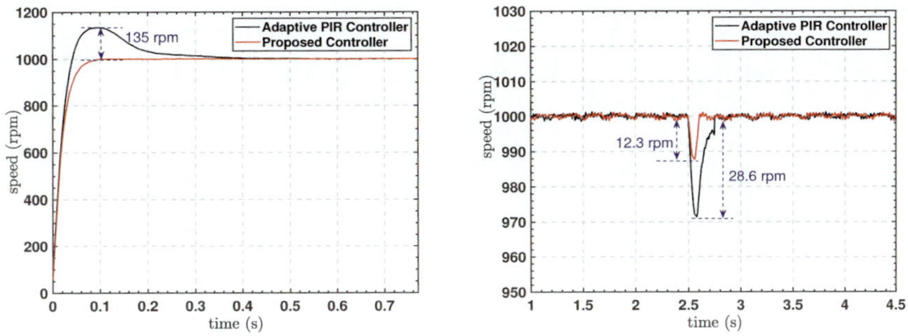

(a) Step-speed-tracking performances. **(b)** Load disturbance rejection performances.

FIGURE 7.9 Step-speed-tracking and load disturbance rejection performances of the proposed controller and adaptive PIR controller.

TABLE 7.1
The Time-Domain Performances Comparison between the Proposed Controller and the Adaptive PIR Controller

	Step-Speed Tracking			Load Disturbance Rejection	
Controller	**Rise Time (s)**	**Overshoot (%)**	**Settling Time (s)**	**Speed Drop (%)**	**Recovery Time (s)**
Proposed	0.07	0	0.07	1.23	0.105
APIR	0.038	13.5	0.25	2.86	0.25

7.4.2 Load Disturbance Rejection Performance

Load disturbance rejection experiments are performed to compare the ability to reject load disturbance, using the proposed controller and adaptive PIR controller. Given the reference speed as 1000 rpm, the load disturbance is imposed on the system when the motor reaches a steady state. The experimental results are shown in Figure 7.9b.

According to Figure 7.9b, the proposed controller has a 12.3 rpm speed drop and the adaptive PIR controller has a 28.6 rpm speed drop. The recovery times of two control system are 0.105 s and 0.25 s, respectively. The time-domain performances of two control systems are summarized in Table 7.1.

7.4.3 Speed Ripple Performance

The speeds of 50, 60, 70, and 80 rpm have been presented in Figures 7.10–7.13. All results of FFT for different speeds are also shown in Figures 7.10–7.13. Table 7.2 shows the calculated total harmonic distortion (THD) and the speed ripple comparisons due to the offset error and scaling error.

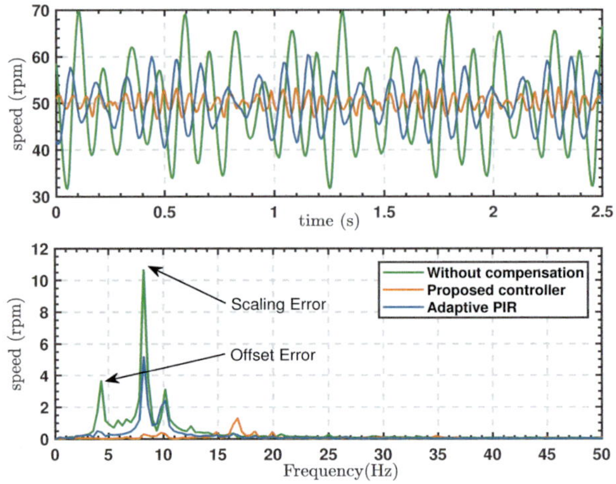

FIGURE 7.10 The steady-state speed performances and FFT of the proposed controller and adaptive PIR controller (50 rpm).

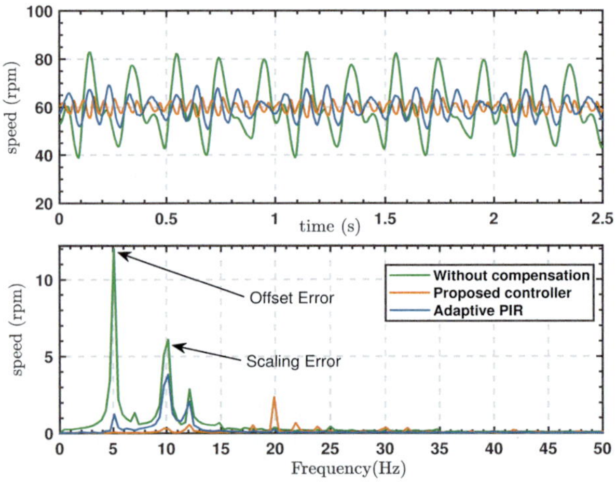

FIGURE 7.11 The steady-state speed performances and FFT of the proposed controller and adaptive PIR controller (60 rpm).

The overall improvement obtained by the proposed controller compared with the adaptive PIR controller is clearly confirmed in time and frequency domains, as well as with the numerically computed values from FFTs as given in Table 7.2. Figures 7.10–7.13 show that the proposed controller can handle speed ripple better due

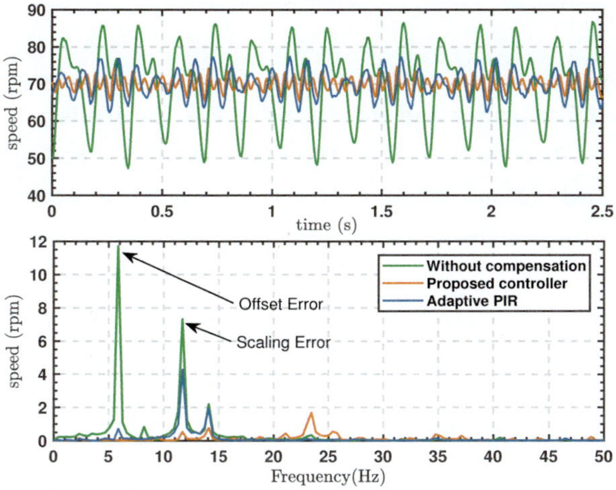

FIGURE 7.12 The steady-state speed performances and FFT of the proposed controller and adaptive PIR controller (70 rpm).

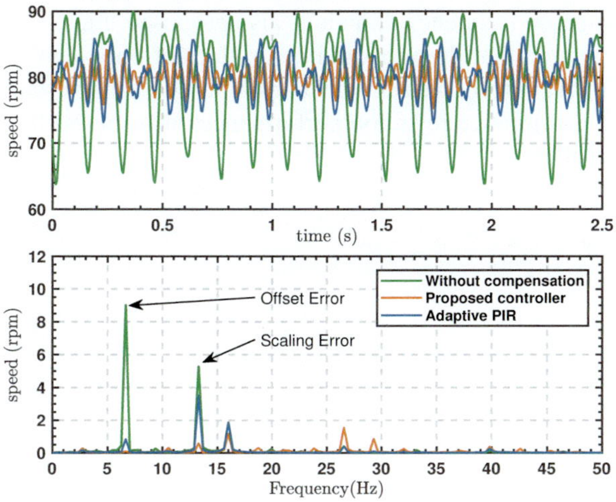

FIGURE 7.13 The steady-state speed performances and FFT of the proposed controller and adaptive PIR controller (80 rpm).

to the current measurement error than the adaptive PIR controller. However, the second harmonic component is not always of lower value. A possible solution would be to include another resonant controller tuned to cancel it.

TABLE 7.2
THD and Current Measurement Error Comparisons of Figures 7.10–7.13 FFTs for the Proposed Controller and the Adaptive PIR Controller.

	THD (%)		Offset Error (rpm)		Scaling Error (rpm)	
Controller	Proposed	Adaptive PIR (s)	Proposed	Adaptive PIR	Proposed	Adaptive PIR
50 rpm	8.3	9.61	0.01654	0.424	0.2876	5.171
60 rpm	2.78	6.05	0.0679	1.264	0.2889	3.837
70 rpm	7.34	7.56	0.03607	0.7211	0.5299	4.28
80 rpm	7.77	14.56	0.0958	0.8378	0.5783	3.514

7.5 CHAPTER SUMMARY

Aiming at the suppression of the speed ripple caused by the current measurement error, this chapter proposed a new control scheme for a PMSM speed servo system. In this method, an observer named as QFR-NESO is proposed, which consists of a NESO and a QFR controller, to deal with load disturbances and periodic disturbances. Then, the proposed controller is applied in the PMSM control experimental system to track the given speed. To facilitate the practical applications, a parameters design method based on frequency domain is also proposed. Finally, the experimental results demonstrated that the proposed controller not only significantly reduces speed ripple due to the current measurement error but also achieves better step-speed tracking and load disturbance rejection performances than the well-known optimal adaptive PIR controller.

8 Resonant-Model-Based ADRC Compensation with Future Work on FO-ADRC

8.1 INTRODUCTION

In Chapter 7, a PMSM speed controller is proposed, including a PD controller, a quasi-fractional-resonant controller and a normalized extended state observer (NESO) to deal with the periodic disturbances in the speed loop. However, the periodic disturbances also widely exist in the current loop, such as flux harmonics [118, 119], and dead time effects. Thus, in this chapter, all the disturbances/uncertainties are taken into account in the controller design for the dual loop to constrict negative effects.

8.2 RMESO-BASED ADRC IN THE CURRENT AND SPEED LOOP

8.2.1 Current-Loop Controller Design

The current-loop control plot is shown in Figure 8.1. The proposed controller includes a resonant-model-based extended state observer (RMESO) and a feedback controller.

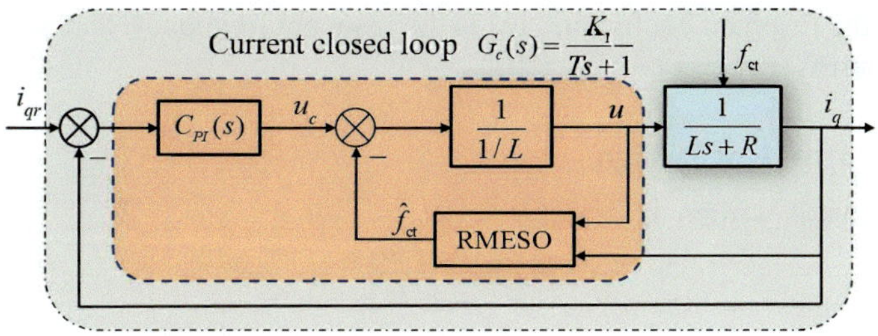

FIGURE 8.1 Block diagram of a PMSM current-loop control system based on the proposed current controller.

8.2.1.1 RMESO Design

The state-space form of the current dynamic is

$$\begin{bmatrix} \dot{i}_q \\ \dot{f}_{ct} \end{bmatrix} = \begin{bmatrix} -\frac{R}{L} & 1 \\ 0 & 0 \end{bmatrix} \begin{bmatrix} i_q \\ f_{ct} \end{bmatrix} + \begin{bmatrix} \frac{1}{L} \\ 0 \end{bmatrix} u_q + \begin{bmatrix} 0 \\ 1 \end{bmatrix} \dot{f}_{ct}. \tag{8.1}$$

DOI: 10.1201/9781003397632-11

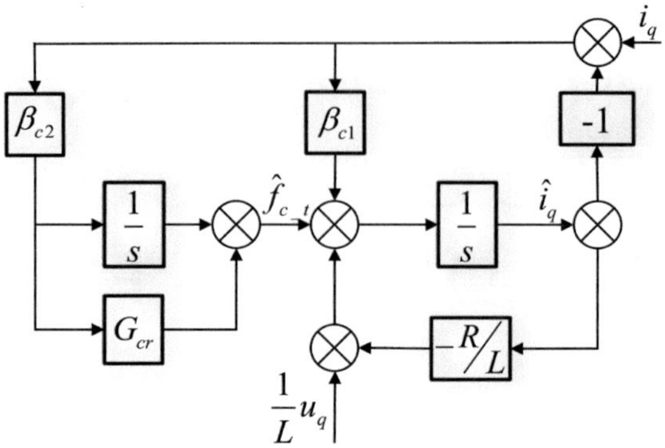

FIGURE 8.2 Block diagram of RMESO.

To estimate the total disturbance f_{ct}, an RMESO is designed,

$$\begin{bmatrix} \dot{\hat{i}}_q \\ \dot{\hat{f}}_{\text{ct}} \end{bmatrix} = \begin{bmatrix} -\frac{R}{L} & 1 \\ 0 & 0 \end{bmatrix} \begin{bmatrix} \hat{i}_q \\ \hat{f}_{\text{ct}} \end{bmatrix} + \begin{bmatrix} \frac{1}{L} \\ 0 \end{bmatrix} u_q + \boldsymbol{L_c}[i_q - \hat{i}_q]. \tag{8.2}$$

where $\boldsymbol{L_c} = [\beta_{\text{c1}}\ \beta_{\text{c2}}(1+G_{\text{cr}}(s)s)]^{\text{T}}$, β_{c1} and β_{c1} are the observer gains, $G_{\text{cr}}(s) = G_{\text{cr1}}(s)+G_{\text{cr2}}(s)+G_{\text{cr3}}(s)\cdots$. $G_{\text{cr}i}(s)$ is a resonant controller [120], which can deal with periodic disturbances. The proposed RMESO is presented in Figure 8.2. The transfer function of the resonant controller is

$$G_{\text{cr}i}(s) = k_{ri}\frac{s\cos\phi - \omega_{ri}\sin\phi}{s^2+\omega_{ri}^2}, \tag{8.3}$$

where k_{ri} is the resonant coefficient, ω_{ri} is the resonant frequency, and ϕ is the phase adjustment term.

8.2.1.2 Control Law Design

The control law is written as

$$u_q = \frac{u_c - \hat{f}_{\text{ct}}}{1/L}, \tag{8.4}$$

where u_c is the output of the proportional integral (PI) controller. The transfer function of the PI controller can be expressed as

$$C_{PI}(s) = K_{cp}(1+\frac{K_{ci}}{s}), \tag{8.5}$$

where K_{cp} and K_{ci} are proportional and integral gains, respectively.

8.2.2 Speed-Loop Controller Design

The speed-loop control plot is shown in Figure 8.3.

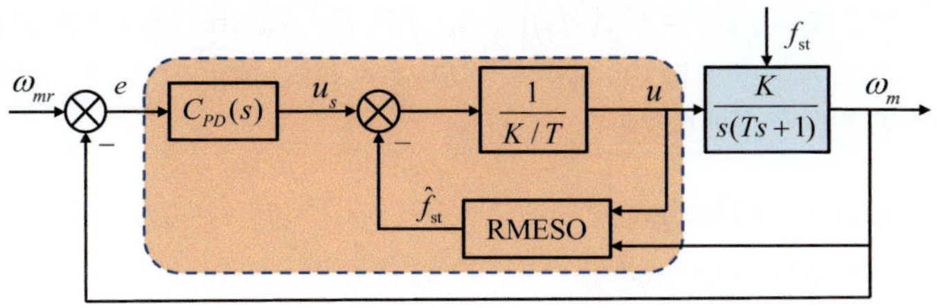

FIGURE 8.3 Block diagram of the PMSM speed-loop control system based on the proposed speed controller.

8.2.2.1 RMESO Design

The speed-control plant consists of the current closed-loop and mechanical dynamic model. The current closed-loop transfer function can be expressed as

$$G_c(s) = \frac{1}{Ts+1}, \tag{8.6}$$

where $T = \frac{1}{K_p}$ and the derivation is given in Sec. 8.3.1.

Combining (8.6) and the mechanical dynamic model, the speed control plant can be obtained as

$$P_s(s) = \frac{Y(s)}{U(s)} = \frac{K}{s(Ts+1)}, \tag{8.7}$$

where $K = 1.5\psi_f p/J$, Y and U represent the motor speed and the output of speed controller, respectively.

The differential equation form of (8.7) can be obtained as

$$\ddot{y} = -\frac{1}{T}\dot{y} + \frac{K}{T}u - f_{st}. \tag{8.8}$$

Define $\boldsymbol{z} = [z_1\ z_2\ z_3]^{\mathrm{T}} = [y\ \dot{y}\ f_{st}]^{\mathrm{T}}$. Assume that f_{st} is differentiable and let $\dot{f}_{st} = h$. Then (8.8) can be rewritten as

$$\begin{aligned} \dot{\boldsymbol{z}} &= \boldsymbol{A}\boldsymbol{z} + \boldsymbol{B}u + \boldsymbol{E}h \\ y &= \boldsymbol{C}\boldsymbol{z}, \end{aligned} \tag{8.9}$$

where $\boldsymbol{A} = \begin{bmatrix} 0 & 1 & 0 \\ 0 & -1/T & 1 \\ 0 & 0 & 0 \end{bmatrix}$, $\boldsymbol{B} = \begin{bmatrix} 0 \\ K/L \\ 0 \end{bmatrix}$, $\boldsymbol{E} = \begin{bmatrix} 0 \\ 0 \\ 1 \end{bmatrix}$, $\boldsymbol{C} = \begin{bmatrix} 1 & 0 & 0 \end{bmatrix}$.

Consequently, the RMESO for the speed-control plant is designed according (8.9),

$$\begin{aligned} \dot{\hat{\boldsymbol{z}}} &= \boldsymbol{A}\hat{\boldsymbol{z}} + \boldsymbol{B}u + \boldsymbol{L}_s(y - \hat{y}) \\ \hat{y} &= \boldsymbol{C}\hat{\boldsymbol{z}}, \end{aligned} \tag{8.10}$$

where $\boldsymbol{L_s} = [\beta_{s1}\ \beta_{s2}\ \beta_{s3}(1+G_{sr}(s)s)]^{\mathrm{T}}$, $\beta_{s1}, \beta_{s2}, \beta_{s3}$ are the observer gains, $G_{sr}(s)$ are the resonant controller in the speed loop, $\hat{\boldsymbol{z}} = [\hat{z}_1\ \hat{z}_2\ \hat{z}_3]^{\mathrm{T}}$, $\hat{z}_1$ approximates y, $\hat{z}_2$ approximates $\dot{y}$, and $\hat{z}_3$ tracks f_{st}.

8.2.2.2 Control Law Design

The control law can be designed as

$$u = \frac{u_s - \hat{z}_3}{K/L}, \tag{8.11}$$

where u_s is the output of the proportional derivative (PD) controller. The transfer function of the PD controller is

$$C_{PD}(s) = K_{sp} + K_{sd}s, \tag{8.12}$$

where K_{sp} and K_{sd} are proportional and differential gains, respectively.

8.3 DECOUPLING CONTROL VERIFICATION, DISTURBANCE REJECTION ANALYSIS, AND PARAMETERS TUNING

8.3.1 Decoupling Control Verification

8.3.1.1 For the Inner Current Loop

According to (8.2), $F_{ct}(s)$ can be calculated with the Laplace transform:

$$F_{ct}(s) = \frac{\beta_{c2}(1+G_{cr}(s)s)[(s+\frac{R}{L})i_q - \frac{1}{L}u_q]}{s^2+(\frac{R}{L}+\beta_{c1})s+\beta_{c2}(1+G_{cr}(s)s)}. \tag{8.13}$$

Next, combining (8.4) and (8.13), the transfer function from u_c to i_q can be expressed as

$$G_{cp}(s) = \frac{I_q(s)}{U_c(s)} = \frac{P_c(s)F(s)}{\frac{1}{L}F(s)+\beta_{c2}(1+G_{cr}(s)s)(P_c(s)(s+\frac{R}{L})-\frac{1}{L})}, \tag{8.14}$$

where $P_c(s) = \frac{1}{Ls+R}$ is the transfer function of the electromagnetic dynamic, $F(s) = s^2+(\frac{R}{L}+\beta_{c1})s+\beta_{c2}(1+G_{cr}(s)s)$.

Substituting $P_c(s) = \frac{1}{Ls+R}$ into (8.14) yields

$$G_{cp}(s) = \frac{1}{s+\frac{R}{L}}. \tag{8.15}$$

Clearly, the equivalent transfer function from u_c to i_q through the designed RMESO is unrelated to the resonant controller, and the block diagram of the current control system based on the proposed current controller can be equivalent to Figure 8.4. The

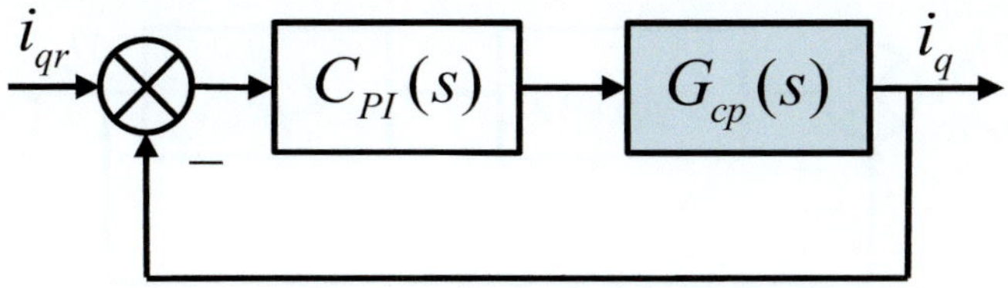

FIGURE 8.4 Equivalent block diagram of the PMSM current-loop control system.

servo performance is solely determined by feedback controller. Thus, the open-loop and closed-loop transfer functions of the inner current-loop control system can be expressed as

$$G_{\mathrm{oc}}(s) = C_{PI}(s)G_{cp}(s), \tag{8.16}$$

$$G_{\mathrm{c}}(s) = \frac{C_{PI}(s)G_{cp}(s)}{1+C_{PI}(s)G_{cp}(s)}. \tag{8.17}$$

8.3.1.2 For the Outer Speed Loop

According to (8.9), $\hat{Z}_3(s)$ can be calculated with the Laplace transform,

$$\hat{Z}_3(s) = \frac{\beta_{\mathrm{s}3}(1+G_{\mathrm{sr}}(s)s)\left(\left(s^2+\frac{1}{T}s\right)Y(s)-\frac{K}{T}U(s)\right)}{E(s)}, \tag{8.18}$$

where $E(s) = s^3 + \left(\frac{1}{T}+\beta_{\mathrm{s}1}\right)s^2 + \left(\frac{1}{T}\beta_{\mathrm{s}1}+\beta_{\mathrm{s}2}\right)s + \beta_{\mathrm{s}3}(1+G_{\mathrm{sr}}(s)s)$.

Then, the transfer function from u_s to ω_m can be represented as

$$G_{sp}(s) = \frac{\Omega_m(s)}{U_s(s)} = \frac{P_s(s)E(s)}{\frac{K}{T}E(s)+\beta_{\mathrm{s}3}(1+G_{\mathrm{sr}}(s)s)(P_s(s)(s^2+\frac{1}{T}s-\frac{K}{T})}. \tag{8.19}$$

Substituting (8.7) into (8.19) yields

$$G_{\mathrm{sp}}(s) = \frac{1}{s(s+\frac{1}{T})}. \tag{8.20}$$

Similarly, the equivalent transfer function from u_s to ω_m through the designed RMESO is nothing with the resonant controller, and the block diagram of the speed control system can be equated to Figure 8.5. The servo performance is only determined by the feedback controller $C_{PD}(s)$. Thus, the open-loop and closed-loop transfer functions of the speed control system can be expressed as

$$G_{\mathrm{os}}(s) = C_{PD}(s)G_{\mathrm{sp}}(s), \tag{8.21}$$

$$G_{\mathrm{cs}}(s) = \frac{C_{PD}(s)G_{\mathrm{sp}}(s)}{1+C_{PD}(s)G_{\mathrm{sp}}(s)}. \tag{8.22}$$

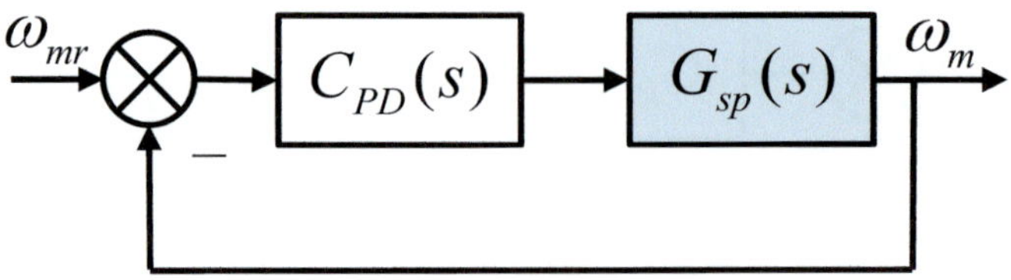

FIGURE 8.5 Equivalent block diagram of the PMSM speed-loop control system.

In summary, both the proposed control schemes in the current loop and speed loop can achieve servo/regulation decoupling control.

8.3.2 Disturbance Rejection Analysis

8.3.2.1 For the Inner Current Loop

The transfer function from total disturbance f_{ct} to the speed-tracking error e is analyzed. Define $i_{qe} = i_q - \hat{i}_q$, $f_{cte} = f_{ct} - \hat{f}_{ct}$; the estimation errors can be obtained:

$$\begin{bmatrix} \dot{i}_{qe} \\ \dot{f}_{cte} \end{bmatrix} = \begin{bmatrix} -\beta_{c1} - \frac{R}{L} & 1 \\ -\beta_{c2}(1+G_{cr}(s)s) & 0 \end{bmatrix} \begin{bmatrix} i_{qe} \\ f_{cte} \end{bmatrix} + \begin{bmatrix} 0 \\ 1 \end{bmatrix} \dot{f}_{ct}. \tag{8.23}$$

Then, the transfer function of total disturbance estimation error can be expressed as

$$G_{ce}(s) = \frac{F_{cte}(s)}{F_{ct}(s)} = \frac{s^2 + (\beta_{c1} + \frac{R}{L})s}{s^2 + (\beta_{c1} + \frac{R}{L})s + \beta_{c2}(1+G_{cr}(s)s)}. \tag{8.24}$$

The dynamic model of the q-axis can be expressed as

$$\dot{i}_q = -\frac{R}{L} i_q + \frac{1}{L}(u_q + \frac{f_{ct}}{1/L}). \tag{8.25}$$

Combining (8.4) and (8.25) yields

$$i_q = P_c(s)\frac{u_c - \hat{f}_{ct}}{1/L} - \frac{f_{ct}}{1/L} = G_{cp}(s)(u_c + f_{cte}). \tag{8.26}$$

According to (8.24) and (8.26), the equivalent current control block diagram can be represented as shown in Figure 8.6. Therefore, the sensitive transfer function of total disturbance f_{ct} to i_q tracking error can be calculated:

$$E_c(s) = \frac{E_{iq}(s)}{F_{ct}(s)} = \frac{G_{ce}(s)G_{cp}(s)}{1 + C_{PI}(s)G_{cp}(s)}. \tag{8.27}$$

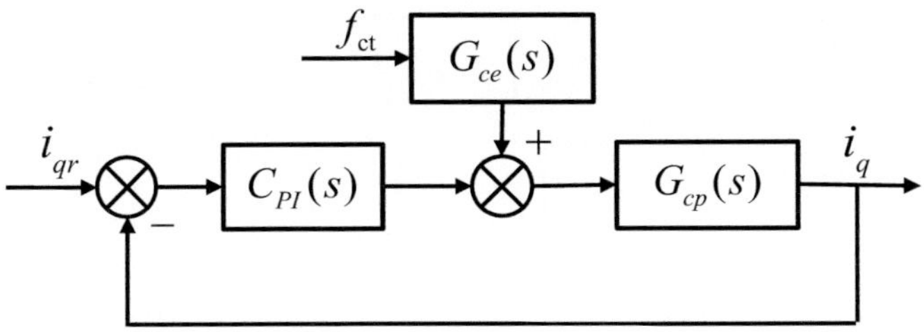

FIGURE 8.6 Equivalent disturbance rejection block diagram of the PMSM current-loop control system.

8.3.2.2 For the Outer Speed Loop

Define $z_1 - \hat{z}_1 = e_{z1}$, $z_2 - \hat{z}_2 = e_{z2}$ and $z_3 - \hat{z}_3 = e_{z3}$. Subtracting (8.10) from (8.9), the estimation errors can be obtained

$$\begin{bmatrix} \dot{e}_{z1} \\ \dot{e}_{z2} \\ \dot{e}_{z3} \end{bmatrix} = \begin{bmatrix} -\beta_{s1} & 1 & 0 \\ -\beta_{s2} & -1/T & 1 \\ -\beta_{s3}(1+G_{sr}(s)s) & 0 & 0 \end{bmatrix} \begin{bmatrix} e_{z1} \\ e_{z2} \\ e_{z3} \end{bmatrix} + \begin{bmatrix} 0 \\ 0 \\ 1 \end{bmatrix} h. \tag{8.28}$$

Next, the transfer function of total disturbance estimation error can be obtained:

$$G_{se}(s) = \frac{E_{z3}(s)}{F_{st}} = \frac{s^3 + \left(\beta_{s1} + \frac{1}{T}\right)s^2 + \left(\beta_{s2} + \frac{1}{T}\beta_{s1}\right)s}{s^3 + \left(\beta_{s1} + \frac{1}{T}\right)s^2 + \left(\beta_{s2} + \frac{1}{T}\beta_{s1}\right)s + \beta_3(1+G_{sr}(s)s)}. \tag{8.29}$$

Similarly, the equivalent speed control block diagram can be expressed as shown in Figure 8.7 and the sensitivity transfer function of disturbance f_{st} to ω tracking error can also be obtained:

$$E_s(s) = \frac{E_{\omega m}(s)}{F_{st}(s)} = \frac{G_{se}(s)G_{sp}(s)}{1 + C_{PD}(s)G_{sp}(s)}. \tag{8.30}$$

8.3.3 Parameters Tuning

8.3.3.1 For the Inner Current Loop

1. K_{cp}, K_{ci}: The parameters K_{ci} of the PI controller in the inner current loop can be obtained through the "cancellation method" for the equivalent plant (8.15) [121], which is commonly used for the current-loop PI controller design, and yields $K_{ci} = \frac{R}{L}$. K_{cp} can be determined with the gain crossover frequency $\omega_{cgc} = 400\pi$ rad/s according to the equation $K_{cp} = \omega_{cgc}$.

2. β_{c1}, β_{c2}: The tuning of the observer gain vector guarantees the idea that place all the poles of MRESO to the same location $-\omega_{co}$, named as observer bandwidth. Thus, $\beta_{c1} = 2\omega_{co} - \frac{R}{L}$, $\beta_{c2} = \omega_{co}^2$. The Bode plots of the total disturbance

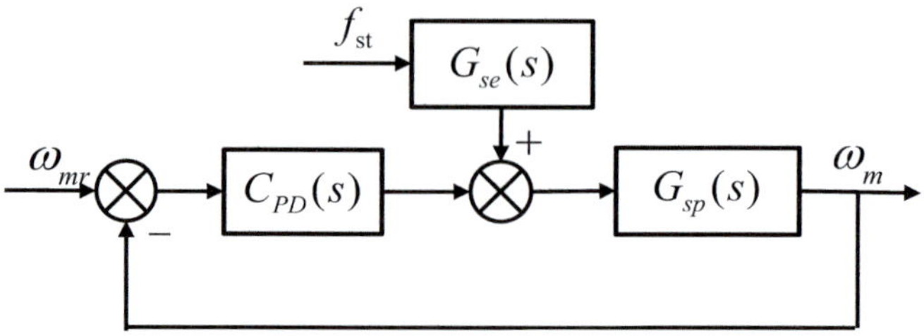

FIGURE 8.7 Equivalent disturbance rejection block diagram of the PMSM speed-loop control system.

estimation error $G_{se}(s)$ with the different bandwidth of RMESO are shown in Figure 8.8a. The magnitude is decreased gradually with the increase of ω_{co}. Theoretically, the disturbance rejection ability increases with increasing ω_{co}. However, the bandwidth is limited by sensor noises. Thus, ω_{co} is set as 2000 rad/s for the experiment in this chapter.

3. ϕ_{ci}, k_{rci}, ω_{rci}: In practical applications, one should determine which factor causes the larger disturbance to the system. Then, ω_{rci} equals to the larger disturbance frequencies. The Bode plots of $G_{se}(s)$ with different k_{rci} are shown in Figure 8.8b. Obviously, with the increasing of k_{rci}, the slot at the resonant frequencies are deeper, which means the ability improved of compensating periodic disturbances. Additionally, it can be observed that the magnitudes near the resonant frequencies spike, which will amplify disturbances at other frequencies. As k_{rci} increases, the magnitudes of the spikes at lower frequencies becomes smaller, but the magnitudes of the spikes at higher frequencies becomes larger. Therefore, the value of k_{rc1} not only affects the response at the resonant frequencies but also affects the response at near frequencies. A comprehensive consideration has been made to choose $k_{rc1} = 0.1$. The Bode plots of $G_{se}(s)$ with different ϕ_{ci} are shown in Figure 8.8c. It can be seen that the spikes can be suppressed with the ϕ_{c1} increasing. Thus, ϕ_{c1} is set as $\frac{50\pi}{180}$ in this chapter.

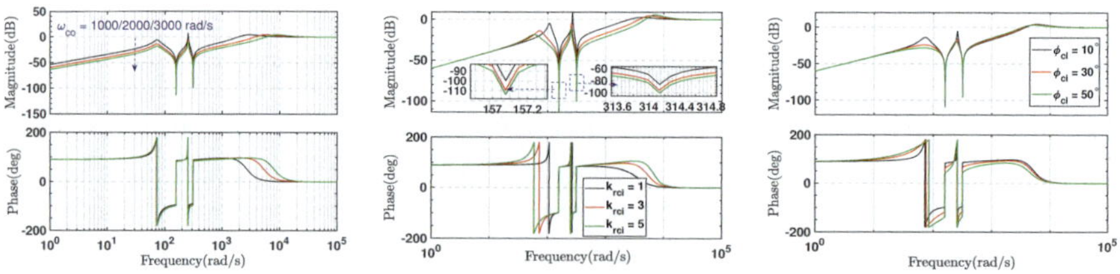

(a) With different ω_{co} **(b)** With different k_{rci} **(c)** With different ϕ_{ci}

FIGURE 8.8 Bode plots of disturbance estimation error.

8.3.3.2 For the Outer Speed Loop

1. K_{sp}, K_{sd}: The parameters K_{sp} and K_{sd} can be determined with the frequency-domain specifications, gain crossover frequency ω_{sgc} and phase margin ϕ_m. Given the frequency-domain specifications ω_{sgc} = 100 rad/s, $\phi_{\text{sm}} = 80°$, controller parameters can be calculated as $K_{sp} = 1.2553 * 10^5$ and $K_{sd} = -119.7950$.

2. β_{s1}, β_{s2}, β_{s3}: Similarly, the observer gain vectors are determined by observer bandwidth ω_{so}, i.e. $\beta_{\text{s1}} = 3\omega_{so} - \frac{1}{T}$, $\beta_{\text{s2}} = 3\omega_{so} - \frac{1}{T}\beta_{\text{s1}}$, and $\beta_{\text{s3}} = \omega_{so}^3$. ω_{so} is set as 500 rad/s for the experiment in this chapter.

3. ϕ_{si}, k_{rsi}, ω_{rsi}: ω_{rsi} equals the disturbance frequencies in the speed loop, which cause larger speed ripples. According to frequency-domain analysis with Bode plots in MATLAB, choose $k_{rs1} = 2$, $k_{rs2} = 2$, $k_{rs3} = 0.5$, $\omega_{rs1} = \frac{50\pi}{180}$, $\omega_{rs1} = \frac{90\pi}{180}$,and $\omega_{rs1} = \frac{50\pi}{10}$.

8.4 EXPERIMENTAL VALIDATION

For fair comparison, the optimal PIR controller [94] and ADRC [122] are designed and tested on dual-loop of PMSM servo system for comparisons to verify the effectiveness of the proposed control strategy. The parameters tuning for the PIR controller is based on frequency-domain specifications gain crossover frequency and phase margin, which are the same as those for the proposed controller to keep a fair comparison. The parameters are designed as $k_p = 3.0159$ and $k_i = 190.9583$, and the resonant controller is the same as the one in the proposed controller. ADRC has two parameters: observer bandwidth and controller bandwidth. The observer bandwidth parameters in current loop and speed loop are the same as the values in our proposed controller, i.e. ω_{co} and ω_{so}, respectively. The controller bandwidth parameters in the current loop and speed loop are the same as the gain crossover frequencies ω_{cgc} and ω_{sgc}, respectively. All the parameters in the PIR and ADRC are designed with the same frequency-domain specifications, thus ensuring the fairness of the comparison.

Three different experiments are carried out, which are given as follows.

1. The first one is to evaluate the step-speed tracking performance of the proposed control strategy.

The reference speed is set as a step change signal of 0 to $\omega_{mr} = \frac{2*\pi*500\text{rpm}}{60}$ rad/s= 52.36 rad/s. Figure 8.9a shows the experimental results of the speed tracking with three different control methods, respectively.

As shown in Figure 8.9a, when the reference speed ω_{mr} is set as a step change signal of 0 to 52.36 rad/s (black dotted line), the actual speed using the PID controller (blue solid line) has a 4.3 rad/s (8%) overshoot and the settling time is 0.145 s. The yellow solid line shows the actual speed response using the ADRC in Figure 8.9a. The overshoot is 16.66 rad/s (31.8%) and the settling time is 0.46 s. Obviously, the designed ADRC control system has a worse performance on step-speed tracking. The analysis is as follows: The effect of equivalent to cascaded integrator by ESO is related to

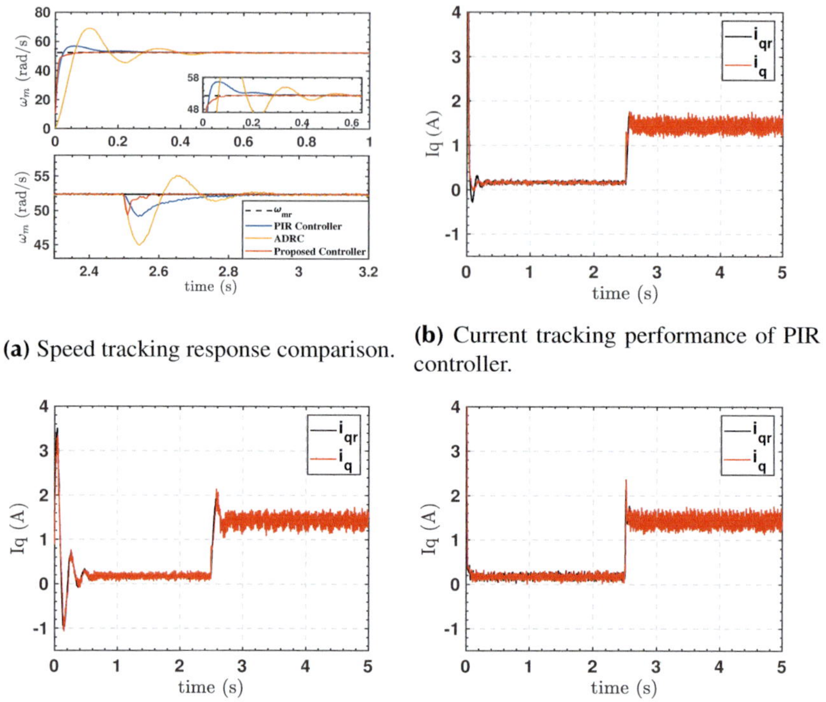

(**a**) Speed tracking response comparison.

(**b**) Current tracking performance of PIR controller.

(**c**) q-axis current tracking performance of ADRC.

(**d**) q-axis current tracking performance of proposed controller.

FIGURE 8.9 Speed responses and current responses for three control systems.

the bandwidth of ESO [54]. In this comparison, the bandwidth of ESO is the same with the bandwidth of the proposed RMESO, keeping fair comparison. Indeed, the performance improves if bandwidth increases for ESO in ADRC. However, the noise will also increase. The actual speed using the proposed controller (red solid line) has no overshoot and the settling time is 0.055 s. It can be clearly seen that the proposed controller achieves the best step-speed tracking performance with no overshoot and obviously shorter settling time than those of the PID controller and ADRC. The settling time represents the interval from the moment when the reference speed is given to that when the motor speed reaches within the error tolerant range, within $\pm 2\%$ of the reference speed.

2. The second experiment is to verify the step load torque rejection performance of the proposed control strategy.

Injecting the load torque when the motor speed becomes stable at 2.5 s, Figure 8.9a shows the experimental comparison results using three controllers. It can be seen in Figure 8.9a that the designed PID controller has a 3.22 rad/s speed drop and the recovery time is 0.14 s. The designed ADRC achieves a 7.48 rad/s speed drop and the recovery time is 0.265 s. Comparing with the designed PID and ADRC control systems, the speed drop of the proposed control system is reduced by 6.8% (from 3.22 rad/s to 3.0 rad/s) and 59.89% (from 7.48 rad/s to 3.0 rad/s), respectively. In terms of recovery time, that of the proposed control system is reduced by 85.71% (0.14 s to 0.02 s) and 92.45% (0.265 s to 0.02 s). Thus, the proposed controller has a better load torque rejection performance than those of PID and ADRC.

3. The third one is to verify the dynamic response performance and periodic disturbance rejection performance of the proposed control strategy at low speed.

In this experiment, the reference speed is set as $\omega_{mr} = \frac{2*\pi*50\text{rpm}}{60} = 5.236$ rad/s. Figure 8.10 shows the q-axis current responses, and Figure 8.11a shows the experimental comparison results (dynamic response performance and steady-state response performance) with three different control methods. Obviously, harsh speed ripples caused at low speed can deteriorate the whole performance.

Firstly, it can be seen from the dynamic response in Figure 8.11a that the proposed control strategy achieves much better performance than others with no overshoot.

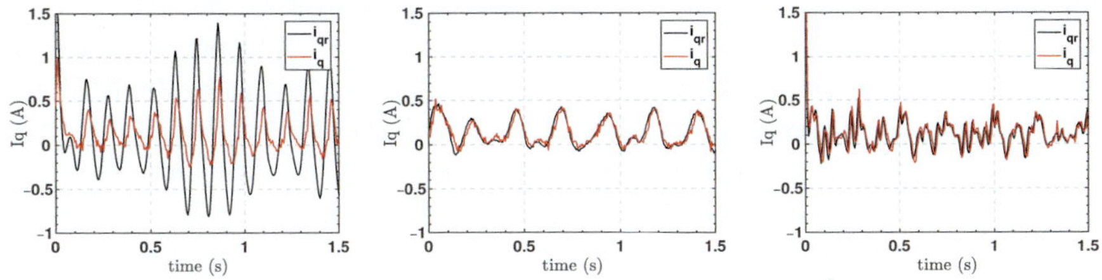

(a) PIR controller. **(b)** ADRC. **(c)** Proposed controller.

FIGURE 8.10 q-axis current responses of three control systems.

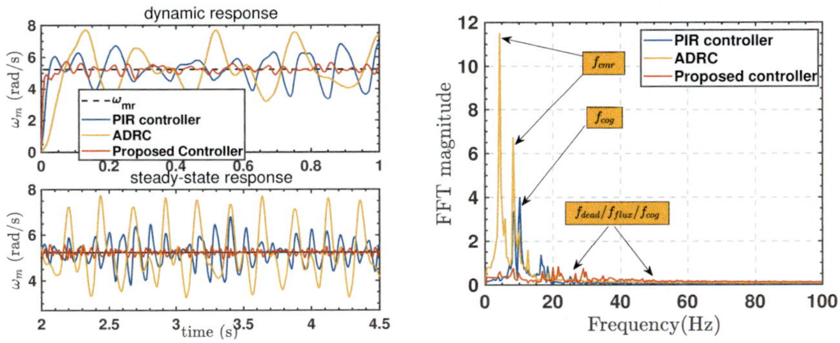

(a) Speed-tracking dynamic response. **(b)** FFT results of steady-state responses.

FIGURE 8.11 Speed responses and FFT of three control systems.

The PIR controller achieves a little overshoot, and the ADRC achieves the biggest overshoot. Then, according to the steady-state response in Figure 8.11a, the speed fluctuations using the PIR controller, ADRC, and the proposed controller are ±1.565 rad/s, ±2.402 rad/s, and ±0.293 rad/s, respectively. Obviously, the proposed controller can achieve better dynamic response performance and steady-state response performance than those of ADRC and the PIR controller. To clearly analyze the factors of fluctuations, fast Fourier transformation (FFT) is utilized, and the results are shown in Figure 8.11b. It can be observed that the current measurement error effects f_{cmr}, the cogging error effects f_{cog} and the dead-time effects using the proposed controller are suppressed to a large extent, compared with ADRC and the PIR controller. Figure 8.11b also shows that the ADRC control system has significant speed ripple and is not sufficiently efficient for periodic disturbances. The reason is that ADRC lacks an internal model of the periodic disturbances. It is worth noting that the high-frequency (30–100 Hz) response is worse with the proposed controller. This is because of the "waterbed effect" in the control system. This question can be addressed with the proposed method in [123].

8.5 CONCLUSION

In this chapter, a systematic control framework is proposed for the PMSM dual loop to attenuate both aperiodic and periodic disturbances. The proposed framework effectively suppresses the aperiodic and periodic disturbances by the RMESO. Through mathematical derivation and theoretical analysis, it is verified that the resonant controller does not affect the open-loop frequency characteristics. Thus, the proposed control framework achieves decoupling control, adjusting the parameters of the RMESO to improve the regulation performance without affecting the servo performance. The disturbance rejection frequency-domain analysis and parameters tuning method are given. Compared with the optimal PIR controller and ADRC, the proposed controller not only largely improves step-speed tracking and load disturbance rejection performances but also significantly reduces speed ripple due to extensive disturbances at low speed.

Part IV

Fractional Order PID and ADR Control

9 Optimal Fractional Order PD and ADR Control

9.1 INTRODUCTION

Due to the fact that, the fractional order proportional derivative (FOPD) controller has been proven to achieve better tracking performance with less overshoot and faster response than traditional integer order proportional derivative controller [11, 15]. Fractional active disturbance rejection control (FADRC) combined with the fractional order extended state observer (ESO) (FOESO) and FOPD controller was proposed by Reference [88] to enhance the control performance for a fractional order system (FOS), where the FADRC stability and frequency-domain characteristics were analyzed. However, the proposed FADRC strictly constrains the orders of FOPD and FOESO corresponding to the order of the FOS. A fractional order active disturbance rejection control (FOADRC) strategy including a nonlinear ESO to achieve precise trajectory tracking performance for a newly designed linear motor was presented in Reference [89]. However, the parameters tuning rule was not presented in Reference [88, 89]. Chen [54] proposed an FOADRC strategy combined with the FOESO and a simple proportional controller. The tuning approach based on frequency-domain specifications was presented for the proposed FOADRC and traditional integer order ADRC. However, the analytical design method based on frequency-domain specifications for ADRC-type controllers has not been studied, which is crucial for industrial applications.

In this chapter, aiming to realize satisfied tracking and anti-load disturbance performance for the permanent magnet synchronous motor (PMSM) speed servo system, a FOADRC combining the FOPD controller and ESO is proposed. The total disturbances are estimated and compensated by the ESO; the FOPD controller achieves optimal tracking performance. A FOADRC design strategy is proposed for satisfying user-specified frequency-domain indexes and achieving a time-domain performance index.

9.2 THE STRUCTURE OF THE FOPD AND ADR CONTROL

In this chapter, a FOADRC with a FOPD controller is proposed.

The differential equation of PMSM speed plant considering the load disturbance can be expressed as

$$\ddot{y} = bu - a\dot{y} + d = bu + f, \tag{9.1}$$

where u and y are the input and output, respectively, d is the external disturbance,

DOI: 10.1201/9781003397632-13

and f is the total disturbance, which contains the internal dynamics and external disturbance d.

The FOADRC includes a FOPD controller and an ESO. The ESO of the FOADRC is used to estimate the total disturbance f. Suppose that f is differentiable, $h = \dot{f}$, then the state space form of (9.1) is

$$\begin{cases} \dot{x} = Ax + Bu + Eh \\ y = Cx, \end{cases} \tag{9.2}$$

where

$$\begin{cases} A = \begin{bmatrix} 0 & 1 & 0 \\ 0 & 0 & 1 \\ 0 & 0 & 0 \end{bmatrix}, B = \begin{bmatrix} 0 \\ b \\ 0 \end{bmatrix}, E = \begin{bmatrix} 0 \\ 0 \\ 1 \end{bmatrix}, \\ C = \begin{bmatrix} 1 & 0 & 0 \end{bmatrix}, x = \begin{bmatrix} x_1 \\ x_2 \\ x_3 \end{bmatrix} \end{cases}$$

Then, the ESO is designed for (9.2),

$$\begin{cases} \dot{z} = Az + Bu + L(y - \hat{y}) \\ \hat{y} = Cz, \end{cases} \tag{9.3}$$

where $L = [\beta_1 \quad \beta_2 \quad \beta_3]^{\mathrm{T}}$ is the gain of the ESO $z = [z_1 \quad z_2 \quad z_3]^{\mathrm{T}}$, z_1, z_2, and z_3 are the outputs of the ESO; z_1 and z_2 will estimate y and its derivative, and z_3 will estimate total disturbance f. The estimated total disturbance will be rejected as

$$u = \frac{u_0 - z_3}{b}, \tag{9.4}$$

where u_0 is the output of the FOPD controller.

The FOPD controller is

$$C_{FOPD}(s) = K_p + K_d s^{\mu}, \tag{9.5}$$

where K_p and K_d are the proportional and derivative gains, and $\mu \in (0,2)$ is the derivative order.

9.3 STABILITY BOUNDARY ANALYSIS

Because the stability is the minimal requirement for the control system, it is desirable to determine the complete stabilizing FOPD parameters before controller design.

From (9.3), with Laplace transform,

$$Z_3(s) = \frac{\beta_3 s^2}{s^3 + \beta_1 s^2 + \beta_2 s + \beta_3} Y(s) - \frac{b\beta_3}{s^3 + \beta_1 s^2 + \beta_2 s + \beta_3} U(s), \tag{9.6}$$

where

$$\beta_1 = 3\omega_o, \quad \beta_2 = 3\omega_o^2, \quad \beta_3 = \omega_o^3. \tag{9.7}$$

In this way, all three of the observer poles will be placed at $-\omega_o$, which is the observer bandwidth [14].

According to (9.4) and (9.6), the plant with disturbance compensation is

$$P_c(s) = \frac{Y(s)}{U_0(s)} = \frac{s^3 + \beta_1 s^2 + \beta_2 s + \beta_3}{s^5 + (a+\beta_1)s^4 + (a\beta_1 + \beta_2)s^3 + (a\beta_2 + \beta_3)s^2}, \tag{9.8}$$

where $U_0(s)$ is the Laplace transform of u_0.

In the considered feedback control system as shown in Figure 9.1, $P_c(s)$ is the controlled plant, and $C_{FOPD}(s)$ is the designed FOPD controller. There are four tuning parameters: K_p, K_d, μ, and ω_o.

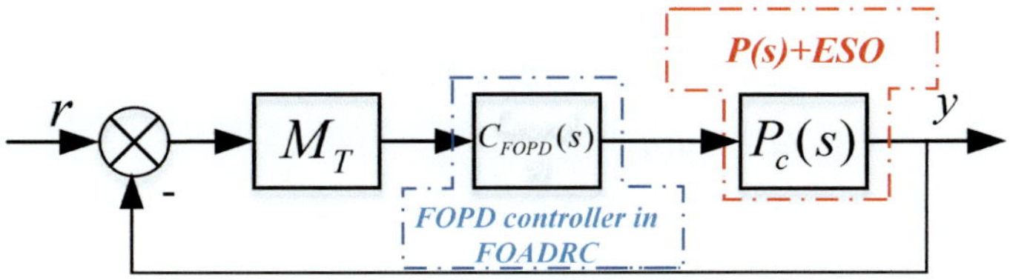

FIGURE 9.1 The PMSM speed closed-loop control system based on FOADRC.

In Figure 9.1, M_T is a gain-phase margin tester,

The closed-loop transfer function of the control system in Figure 9.1 can be obtained as

$$G(s) = \frac{M_T(A,\phi)C_{FOPD}(s)P_c(s)}{1 + M_T(A,\phi)C_{FOPD}(s)P_c(s)}. \tag{9.9}$$

Hence, the characteristic equation of $G(s)$ is

$$\begin{aligned} D(K_p, K_d, \mu, A, \phi; s) = (s^5 + (a+\beta_1)s^4 + (a\beta_1 + \beta_2)s^3 + \\ (a\beta_2 + \beta_3)s^2) + Ae^{-j\phi}(K_p + K_d s^\mu)(s^3 + \beta_1 s^2 + \beta_2 s + \beta_3). \end{aligned} \tag{9.10}$$

The range of the fractional order μ is defined as $\mu \in (0,2)$. The lower limit of ω_o depends on the prespecified gain crossover frequency ω_{gc}, and the upper limit of ω_o is set to 800 rad/s in this chapter. The boundaries of the stable region which are determined by real root boundary (RRB), infinite root boundary (IRB) and complex root boundary (CRB) can be obtained by the D-decomposition method [29, 30].

RRB: The RRB is defined by the equation $D(K_p, K_d, \mu, A, \phi, \omega_o; s = 0) = 0$, so the boundary of K_p is

$$K_p = 0. \tag{9.11}$$

IRB: Due to the relative order of $P_c(s)$ is 2, K_d has no boundary restrictions.

CRB: Putting $j\omega$ for s into (9.10), the CRB can be described by $D(K_p, K_d, \mu, A, \phi, \omega_o; s = j\omega) = 0$, yielding

$$\begin{aligned} D(K_p, K_d, \mu, A, \phi, \omega_o; j\omega) &= ((j\omega)^5 + (a+\beta_1)(j\omega)^4 \\ &+ (a\beta_1 + \beta_2)(j\omega)^3 + (a\beta_2 + \beta_3)(j\omega)^2) \\ +Ae^{-j\phi}(K_p + K_d(j\omega)^{\mu})((j\omega)^3 &+ \beta_1(j\omega)^2 + \beta_2(j\omega) + \beta_3) = 0. \end{aligned} \tag{9.12}$$

The real part and the imaginary part of (9.12) can be obtained,

$$F_1 + A(K_pE_5 + K_dE_6) = 0, \tag{9.13}$$

$$F_2 + A(K_pE_7 + K_dE_8) = 0, \tag{9.14}$$

where

$$\begin{aligned} &F_1 = (a+\beta_1)\omega^4 - (a\beta_2 + \beta_3)\omega^2, \\ &F_2 = \omega^5 - (a\beta_1 + \beta_2)\omega^3, \\ &E_3 = -\omega^3 + \beta_2\omega, \ E_4 = -\beta_1\omega^2 + \beta_3, \\ &E_5 = cos(\phi)E_4 + sin(\phi)E_3, \\ &E_6 = \omega^{\mu} cos(\frac{\mu\phi}{2} - \phi)E_4 - \omega^{\mu} sin(\frac{\mu\pi}{2} - \phi)E_3, \\ &E_7 = cos(\phi)E_3 - sin(\phi)E_4, \\ &E_8 = \omega^{\mu} cos(\frac{\mu\phi}{2} - \phi)E_3 - \omega^{\mu} sin(\frac{\mu\pi}{2} - \phi)E_4. \end{aligned}$$

From (9.13) and (9.14),

$$K_d = \frac{F_2E_5 - F_1E_7}{A(E_6E_7 - E_5E_8)}, \tag{9.15}$$

$$K_p = -\frac{F_2 + AK_dE_8}{E_7}. \tag{9.16}$$

So, given a fixed fractional-order μ and a fixed ω_o, the stable and unstable regions in the parameter plane can be separated according to the RRB and CRB with $A = 1$, $\phi = 0°$.

One example is given as: Choosing $\mu = 0.9$ and $\omega_o = 200$ rad/s, draw the curve of K_p w.r.t. K_d and the line $K_p = 0$ according to CRB and RRB, respectively. Detect the stable region by testing a random point [32] as shown in Figure 9.2a. Sweeping all the $\mu \in (0,2)$, the global stable region can be obtained as shown in Figure 9.2b. Thus, with different ω_o, different global stable regions can be determined.

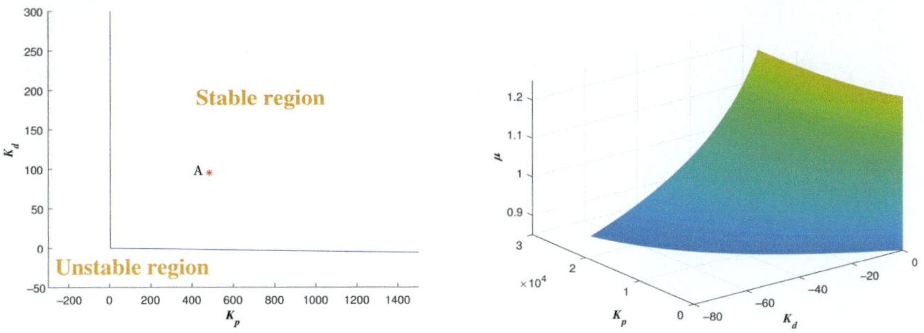

(a) Stable region with $\mu = 0.9$, $\omega_o = 200$ rad/s, and the designed K_p and K_d. **(b)** Global stable region with $\omega_o = 200$ rad/s.

FIGURE 9.2 Stable region ($\phi_m = 80^\circ$, $\omega_{gc} = 50$ rad/s) and global stable region.

9.4 FOADRC/IOADRC DESIGN STRATEGY

9.4.1 Controller Design Specifications

In this chapter, two frequency-domain specifications, gain crossover frequency (ω_{gc}) and phase margin (PM), are applied to tune FOADRC. To achieve an optimal dynamic control performance, a time-domain specification, the integral time absolute error (ITAE), is also applied for the FOADRC design.

9.4.2 The Optimal FOADRC Design

9.4.2.1 The FOADRC Satisfying the Frequency-Domain Specifications

The gain-phase margin tester can provide information for satisfying the given gain margin or phase margin [29]. Setting $A = 1$ and $\phi = \phi_m$, plot the curve according to Eqs. (9.15) and (9.16) with ω increasing from 0. Every point on the above curve fulfills the given phase margin ϕ_m.

The characteristic equation of Eq. (9.9) is

$$1 + M_T(A,\phi)C_{FOPD}(s)P_c(s)|_{s=j\omega} = 0, \tag{9.17}$$

which denotes that the open-loop transfer function $T(s)$ equals to -1,

$$T(s)_{s=j\omega} = M_T(A,\phi)C_{FOPD}(s)P_c(s)|_{s=j\omega} = -1. \tag{9.18}$$

So one can get the magnitude and the phase equations,

$$|M_T(A,\phi)C_{FOPD}(s)P_c(s)|_{s=j\omega} = 1$$
$$\arg[M_T(A,\phi)C_{FOPD}(s)P_c(s)]_{s=j\omega} = -\pi,$$

Thus, all the ω satisfying (9.12) can be treated as the gain crossover frequencies ω_{gc} with $A = 1$ and $\phi = \phi_m$ for the system (9.8) in Figure 9.1. Because Eqs. (9.12) and

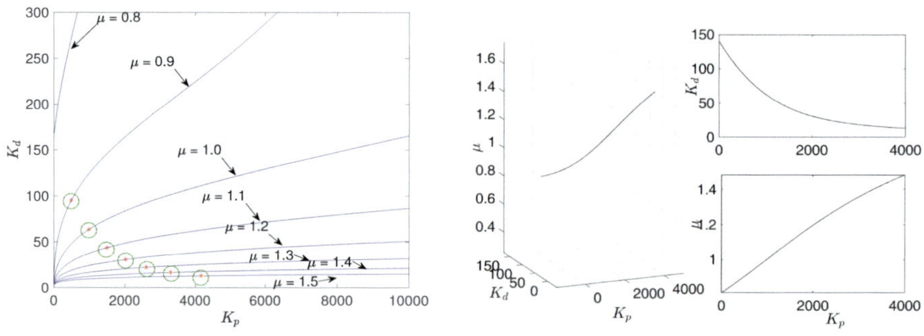

(a) Two-dimensional diagrams.

(b) Three-dimensional diagrams.

FIGURE 9.3 Two-dimensional and three-dimensional diagrams of the designed K_p and K_d with all the $\mu \in (0,2)$.

(9.18) are equivalent, every frequency ω corresponding to the point on the curve of K_p versus K_d in parameter plane (K_p, K_d) can also be treated as ω_{gc}.

Thus, with the specified ω_c, ϕ_m, a fixed μ, and a fixed ω_o, the other two FOADRC parameters K_p and K_d can be determined at a point. Besides, the point should be tested whether it is in the stable region. Then, sweeping over the $\mu \in (0,2)$, a curve in the (K_p, K_d, μ) space can be determined. All the points on this curve can be guaranteed to satisfy the two specifications ω_c and ϕ_m. Then, sweeping all the $\omega_o \in (\omega_{gc}, \omega_{max})$, one can obtain a three-dimensional graph about K_p, K_d, and μ. Every point on this graph corresponding to the parameters (K_p, K_d, μ) of the FOADRC satisfies the prespecified ω_{gc} and ϕ_m.

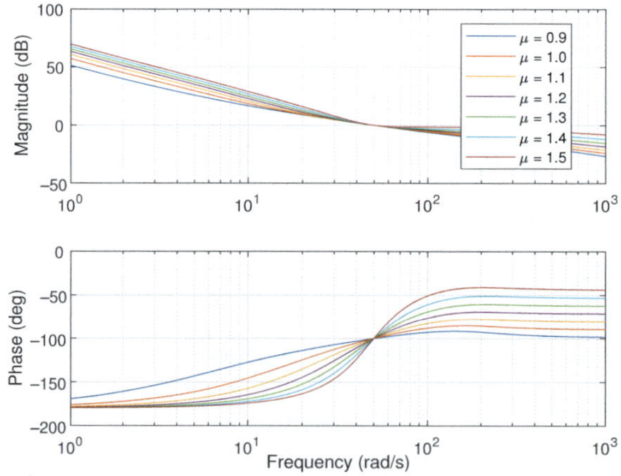

FIGURE 9.4 Bode plots with different μ.

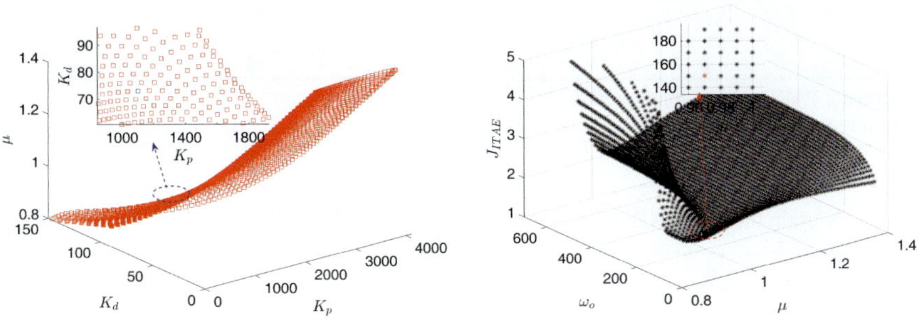

(a) All the K_p, K_d, and μ. **(b)** Corresponding μ, ω_o, and J_{ITAE}.

FIGURE 9.5 All the controller parameters and corresponding J_{ITAE} with $\phi_m = 80°$ and $\omega_{gc} = 50$ rad/s.

One example is given as: With the prespecified $\phi_m = 80°$, $\omega_{gc} = 50$ rad/s and fixed $\mu = 0.9$, and $\omega_o = 200$ rad/s, the point corresponding to the parameters (K_p, K_d) is determined as a red star "A" shown in Figure 9.2a. Sweeping all the μ in $(0,2)$, a series of red stars can be obtained as shown in Figure 9.3a and visualized in three-dimensional parameter space, which is shown in Figure 9.3b. Every point on this curve meets $\omega_{gc} = 50$ rad/s and $\phi_m = 80°$ according to Figure 9.4 with the $\omega_o = 200$ rad/s, simultaneously. Sweeping all the ω_o in (ω_{gc}, 800), one can obtain a three-dimensional scatter plot of $K_p - K_d - \mu$ as shown in Figure 9.5a. Every point corresponding to the parameters satisfies the frequency-domain specifications $\omega_{gc} = 50$ rad/s and $\phi_m = 80°$.

9.4.2.2 The Optimal FOADRC Satisfying Time-Domain Specifications

Given all the parameters of the FOADRC corresponding to the point on the above curve in Figure 9.3b, the step and load-disturbance response simulation are implemented and the ITAE of the trial is calculated in MATLAB/Simulink. Sweeping all ω the in (ω_{gc}, 800), the three-dimensional scatter plot of $\mu - \omega_o - J_{ITAE}$ is shown in Figure 9.5b. According to Figure 9.5b, the smallest J_{min} can be determined, and the parameters $(K_p, K_d, \omega_o, \mu)$ of the optimal FOADRC corresponding to the J_{min} can be obtained.

In summary, the design guidelines are shown in Figure 9.6.

9.4.3 IOADRC Design Strategy

The design strategy for IOADRC satisfying the user-specified frequency specifications ω_{gc} and PM is also given with a short explanation. For a fair comparison with FOADRC, set the same ω_o with FOADRC. Just as analyzed above, the parameters K_p and K_d with $\mu = 1$ can be determined according to Eqs. (9.15) and (9.16), with $\omega = \omega_{gc}$, $\phi = \phi_m$.

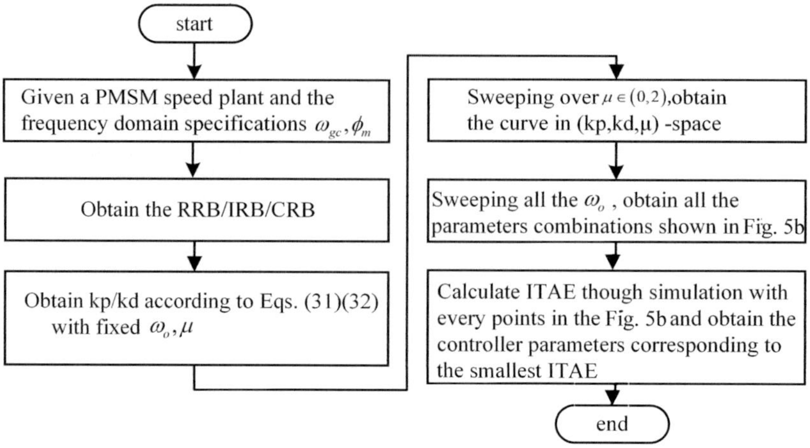

FIGURE 9.6 The design guidelines summary.

9.5 SIMULATION RESULTS

In this section, simulation studies are carried out on a PMSM speed servo system compared with IOADRC and a typical PID controller [125]. The tuning strategy for a typical PID controller is based on the frequency-domain specifications (gain crossover frequency, phase margin, and flat phase constraint).

9.5.1 Tracking Performance

Given the frequency-domain specifications $\omega_{gc} = 10$ rad/s and $\phi_m = 60°$, the optimal FOADRC is obtained applying the above proposed method,

$$K_{p1} = 123.59, K_{d1} = 36.248, \mu = 0.74, \omega_o = 40 \text{ rad/s}. \tag{9.19}$$

For fair comparison, the IOADRC is designed with the same frequency-domain specifications $\omega_{gc} = 10$ rad/s and $\phi_m = 60°$ and the same $\omega_o = 40$ rad/s, so that one can get

$$K_{p2} = 202.703, K_{d2} = 18.282. \tag{9.20}$$

Similarly, the typical PID controller can be designed with the same frequency-domain specifications as

$$C_{PID}(s) = 0.719 + \frac{1.7416}{s} + 0.006s. \tag{9.21}$$

The open-loop Bode plots of the three control systems are presented in Figure 9.7a. Three control systems have the same ω_{gc} and ϕ_m. The closed-loop Bode plots of three control systems are shown in Figure 9.7b. Figure 9.7b shows that the designed

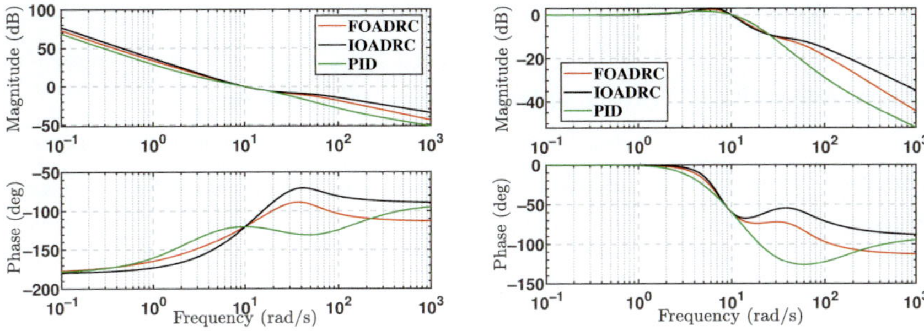

(a) Open-loop Bode diagrams.

(b) Closed-loop Bode diagrams.

FIGURE 9.7 Open-loop and closed-loop Bode diagrams comparison of three control systems.

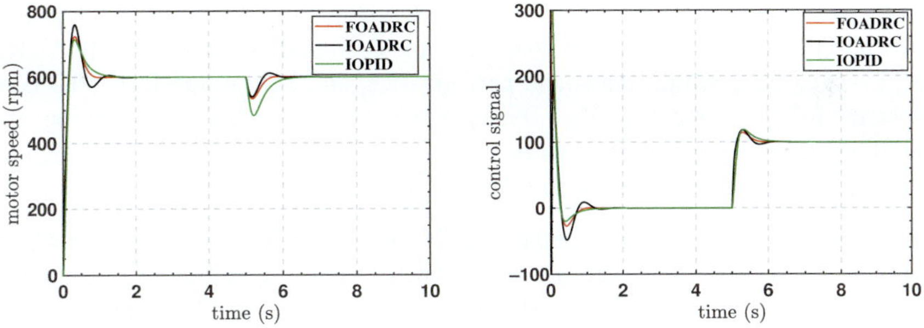

(a) Speed responses.

(b) Control signals.

FIGURE 9.8 The speed response and control signals comparison of three control systems.

IOADRC control system has the biggest resonance peak, which means the IOADRC control system has the biggest overshoot than other control systems. The designed PID control system has the smallest overshoot with the smallest resonance.

Given the reference speed as 600 rpm, PMSM speed responses are performed, using the designed FOADRC, IOADRC, and PID controller. The simulation results are shown in Figures 9.8a and 9.8b. The comparison results are shown in Table 9.1.

Obviously, the designed PID control system has the smallest overshoot and the designed FOADRC control system has the biggest overshoot. The overshoots of three control systems are consistent with the analysis in Figure 9.7b. However, compared to the IOADRC control system, the designed FOADRC control system has shorter settling time (0.7369 s) and smaller overshoot (20.5%); compared to the PID control system, the designed FOADRC control system has a shorter settling time, although the overshoot is bigger. Overall, the designed FOADRC control system achieves the

TABLE 9.1
Comparison Results of Three Control Systems (Simulation)

Controller	Speed Tracking: Overshoot (%)	Speed Tracking: Settling Time (s)	Speed Tracking: J_{ITAE}	Anti-Load Disturbance: Speed Drop (%)	Anti-Load Disturbance: J_{ITAE}
FOADRC	20.5	0.7369	19.2	12.07	110.32
IOADRC	26.4	0.9831	27.18	11.08	124.3
PID	19	0.985	28.05	19.47	301.45

best tracking performance by comparison with J_{ITAE} and can achieve a compromise between overshoot and settling time.

9.5.2 Robustness to External Disturbance

The process sensitivity Bode plots $(P(s)/(1+C(s)P(s)))$ for three control systems are shown in Figure 9.9. Figure 9.9 shows that the designed FOADRC and IOADRC control systems have similar anti-load performance, and both outperform PID control system. Injecting the load disturbance when the motor speed is running, the anti-load disturbance responses of three control systems are performed as shown in Figure 9.8a, and the comparison results are shown in Table 9.1.

As shown in Figure 9.9, the designed FOADRC and IOADRC control systems have the similar speed drop, and both are less than the speed drop of the PID control system. According to Figure 9.8a and Table 9.1, the designed PID control system has the biggest speed drop (19.47%) and recovery time; the speed drop shows that the anti-load disturbance performances of FOADRC and IOADRC control systems

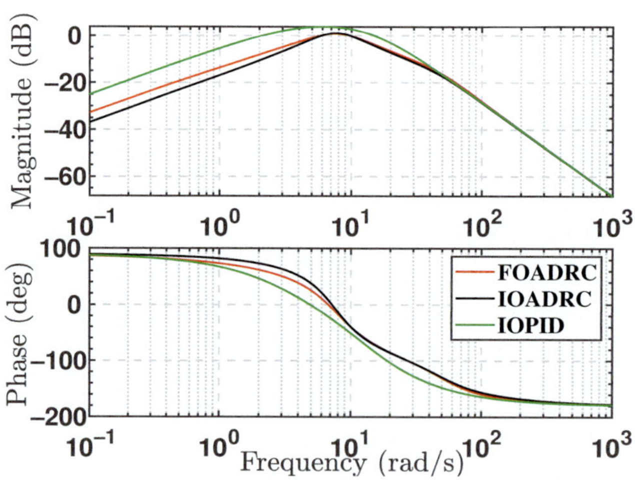

FIGURE 9.9 Sensitivity Bode diagrams of three control systems.

are basically the same and both significantly smaller than the PID control system, which also can be seen clearly from J_{ITAE}. These results are consistent with the analysis from Figure 9.9.

9.6 EXPERIMENT VALIDATION

9.6.1 Speed-Tracking and Load Disturbance Rejection Performance

Setting the reference speed as 600 rpm, PMSM speed step response experiments are performed, using the designed FOADRC, IOADRC, and PID controller. The experimental results are presented in Figure 9.10, and the comparison results are shown in Table 9.2.

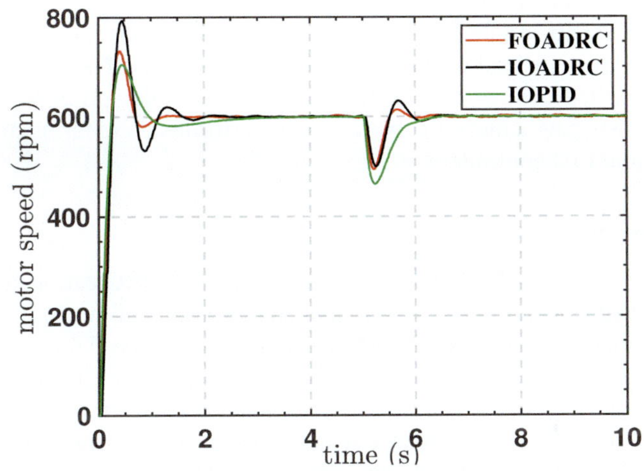

FIGURE 9.10 The speed and anti-load disturbance responses of three control systems (experiment).

TABLE 9.2
Comparison Results of Three Control Systems (Experiment)

	Speed Tracking			Anti-Load Disturbance	
Controller	Overshoot (%)	Settling Time (s)	J_{ITAE}	Speed Drop (%)	J_{ITAE}
FOADRC	22.1	0.985	28.72	17.6	174.814
IOADRC	32.3	1.47	60.58	16.6	210.8931
PID	19	1.94	64.3414	19.47	368.0378

Figure 9.10 shows that the designed PID control system has the smallest overshoot, and the designed IOADRC control system achieves the biggest overshoot, which

is consistent with the analysis from Figure 9.7b and the simulation results in Figure 9.8a. Compared to the IOADRC control system, the designed FOADRC control system has shorter settling time (0.985 s) and smaller overshoot (22.1%); the designed PID control system has the smallest overshoot but the longest settling time. In summary, the FOADRC control system achieves the best tracking performance according to the J_{ITAE} and can achieve a compromised performance in terms of overshoot and settling time between IOADRC and PID control systems.

Injecting the same disturbance, the anti-load disturbance responses of three control systems are performed in Figure 9.10. As can be seen visually from Figure 9.10, IOADRC and FOADRC control systems achieve the similar speed drop and both less than the speed drop of the PID control system. These results are consistent with the analysis from Figure 9.9 and the simulation results in Figure 9.8a. The designed PID control system has the biggest speed drop (19.47%) and recovery time. The IOADRC and FOADRC control systems have basically similar speed drop (16.6% and 17.6%), and the anti-load disturbance performance of IOADRC has a little oscillation. The detailed comparison results are shown in Table 9.2. In summary, the proposed FOADRC can achieve the best control performance in terms of anti-load over IOADRC and PID controllers.

9.6.2 Discussion

Compared to the traditional IOADRC, the proposed FOADRC can achieve better tracking performance with a smaller overshoot, and the value of the J_{ITAE} is smaller. On the other hand, the proposed FOADRC has a little bigger speed drop when inputs load compared to IOADRC. However, recover time is less than IOADRC and has a smaller value of J_{ITAE}. Compared to the traditional PID controller, the proposed FOADRC can achieve better disturbance rejection performance with less speed drop when inputs load. On the other hand, the proposed FOADRC has a bigger overshoot compared to PID controller. However, the settling time of the proposed FOADRC is less than that of the PID controller and has less J_{ITAE}.

9.7 CONCLUSIONS

The chapter proposes a FOADRC with a FOPD controller for the PMSM speed servo system, which is identified through a nonlinear identification method. A stabilization method is presented to obtain the stabilizing FOPD controller for the PMSM speed servo plant via D-decomposition. Furthermore, a synthesis is proposed for the FOADRC design to meet user-specified frequency-domain specifications, i.e., phase margin and gain crossover frequency, and the time-domain specification, i.e., ITAE, simultaneously. The design method is also applicable to IOADRC. The designed optimal FOADRC/IOADRC can get desired control performance satisfying two given frequency-domain specifications and the optimal J_{ITAE}. The simulation and experiment are carried out for the PMSM speed servo, comprised of the traditional IOADRC and PID controller. Based on the simulation and experiment results, the proposed FOADRC can achieve better tracking performance and more robustness to external disturbance. The J_{ITAE} for speed tracking of the designed FOADRC are 52.59% and 55.36% better than the J_{ITAE} of IOADRC and PID controller, respectively. Obviously, the settling time of the designed FOADRC is the shortest (0.7369 s), which is 32.99% and 49.22% less than the settling time of the designed IOADRC and PID controller. Besides, the J_{ITAE} for anti-load disturbance of the designed FOADRC is 17.11% and 52.50% better than the J_{ITAE} of IOADRC and PID controller, respectively.

10 A Two-Degree-of-Freedom Controller Design Satisfying Separation Principle with FOPD and Generalized ESO

10.1 INTRODUCTION

Facing the problem of inherent trade-off on tracking and disturbance rejection, a two-degrees-of-freedom controller has been investigated. In [126], a two-degrees-of-freedom controller including a proportional integral derivative (PID) and a set-point filter was proposed. The PID controller guarantees the regulatory response, and the set-point filter achieves satisfactory servo performance. In [127], a two-degrees-of-freedom controller was proposed and verified that the resulting closed-loop control system is robust with respect to parameter uncertainties and achieved outstanding torque disturbance rejection. The disturbance-observer-based control methods have been proposed and practiced by many researchers and engineers, such as generalized proportional integral observer [128], extended state observer (ESO) [129], and uncertainty and disturbance estimator [130]. However, all these methods cannot fully satisfy the separation principle to achieve optimal control performances for both tracking and anti-load disturbance simultaneously. The above methods still cannot meet the desired optimal set-point tracking and anti-load disturbance rejection simultaneously for the controller design.

The internal model control (IMC) is an easily understandable approach with robustness [131]. The studies showed that an IMC control can achieve optimal tracking performance, but the load disturbance rejection performance is not satisfactory [131, 132]. The sliding mode controller is an effective control method [133]. However, it has not been widely used in industrial applications until now because of the chattering which is caused by switch function [133, 134]. Inverse-model-based control has also been investigated [135, 136]. Two critical problems are the computation of the inverse model and the design of the filter $Q(s)$, which affects the disturbance suppression [137]. Extended state observer, as the core idea of the active disturbance rejection controller, has excellent disturbance rejection effect. Therefore, the ESO is widely used in industrial applications [138–140]. A general active disturbance rejection (GADRC) using plant information was proposed in [141]

DOI: 10.1201/9781003397632-14

with stability analysis. Simulation results proved that better control performance can be achieved compared to conventional ESO [142]. The ADRC research is a hot topic for disturbance rejection control in recently years, with great potential for robustness improvement based on existing literatures. Therefore, the ESO in ADRC is applied in our proposed two-degree-of-freedom control methodology to pursue advanced robustness improvement with separation principle. The fractional order PD (FOPD) controller has been proven to achieve better tracking performance with less overshoot than a traditional integer order PD controller [15, 124, 143]. In this chapter, a two-degrees-of-freedom FOPD-general ESO (FOPD-GESO) controller is proposed. The structure of the FOPD-GESO controller consists of a FOPD controller and a GESO. The GESO simplifies the plant and achieves desired load-disturbance performance, and FOPD achieves outstanding tracking performance, which breaks the inherent trade-off on tracking response and disturbance rejection of the typical PID controller. Moreover, the parameter tuning method of the proposed FOPD-GESO fully meets the separation principle, which means the anti-load disturbance can be adjusted by the parameters of GESO, and the tracking performance can be adjusted by the parameters of FOPD, respectively.

The major purpose of the work in this chapter is to break the inherent trade-off between the set-point tracking and anti-load disturbance of the typical PID controller and solve the difficult problem of ADRC tuning. To summarize, the main contributions of this chapter are as follows: (1) An FOPD-GESO controller is proposed to improve the performance of the typical PMSM speed servo motion system. (2) A systematic design scheme for the FOPD-GESO controller is proposed for the permanent magnet synchronous motor (PMSM) speed servo system with the separation principle, which can also be extended to typical second-order and third-order plants. (3) Because of satisfying the separation principle, *set-point tracking (servo)* and *load-disturbances rejection (regulation)* can be optimized by tuning the FOPD controller and GESO separately, which breaks the inherent trade-off on *servo* and *regulation* of the typical PID control system. (4) Simulation and experiments on the PMSM speed servo system demonstrate that the proposed FOPD-GESO controller outperforms the optimal designed traditional PID controller, integer order ADRC (IOADRC), and FOPID controller.

10.2 CONTROLLER DESIGN ANALYSIS

10.2.1 FOPD-GESO Controller

In [141], the GESO with control plant information is proposed to improve control performance. According to Eq. (9.1), the GESO can be designed as

$$\begin{cases} \begin{bmatrix} \dot{z}_1 \\ \dot{z}_2 \\ \dot{z}_3 \end{bmatrix} = \begin{bmatrix} 0 & 1 & 0 \\ 0 & -a & 1 \\ 0 & 0 & 0 \end{bmatrix} \begin{bmatrix} z_1 \\ z_2 \\ z_3 \end{bmatrix} + \begin{bmatrix} 0 \\ b_0 \\ 0 \end{bmatrix} u + L(y-\hat{y}), \\ \hat{y} = \begin{bmatrix} 1 & 0 & 0 \end{bmatrix} \begin{bmatrix} z_1 \\ z_2 \\ z_3 \end{bmatrix} \end{cases} \tag{10.1}$$

where z_1, z_2, and z_3 are the outputs of GESO, z_1 and z_2 approximate y and its derivatives, z_3 approximates the disturbance f, $b_0 = b$, and L is the observer gain vector:

$$L = [\beta_1 \quad \beta_2 \quad \beta_3]^T. \tag{10.2}$$

The tuning of the observer gain is reduced to one parameter as suggested in [14]: the observer bandwidth ω_o. The idea is to place all the poles of the GESO to the same location $-\omega_o$. The elements of L are obtained as $\beta_1 = 3\omega_o - a, \beta_2 = 3\omega_o^2 - a\beta_1, \beta_3 = \omega_o^3$.

The estimated disturbance can be rejected as

$$u = \frac{u_0 - z_3}{b_0}, \tag{10.3}$$

where u_0 is the output of the FOPD controller.

The FOPD controller can achieve the desired tracking performance. Its transfer function is

$$C_{FOPD}(s) = K_p + K_d s^{\mu}, \tag{10.4}$$

where K_p and K_d are proportional and differential gains, respectively, μ is the fractional order, and $\mu \in (0,2)$. The control structure is shown in Figure 10.1.

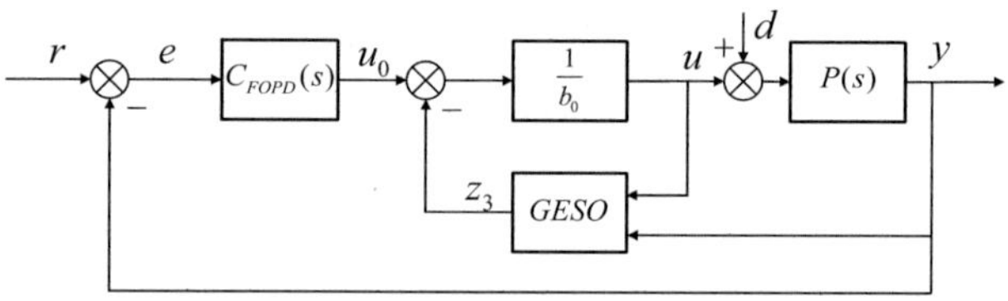

FIGURE 10.1 Feedback control system.

10.2.2 Frequency-Domain Analysis

From Eq. (10.1), through Laplace transform and solving the equations, $Z_3(s)$ can be obtained as

$$Z_3(s) = \frac{\beta_3(s^2 + as)Y(s) - \beta_3 b_0 U(s)}{s^3 + (a+\beta_1)s^2 + (a\beta_1 + \beta_2)s + \beta_3}. \tag{10.5}$$

Then, according to the $P(s) = Y(s)/U(s)$ and Eqs. (10.3) and (10.5), the simplified model, i.e. the transfer function from u_0 to y, can be shown as

$$P_s(s) = \frac{Y(s)}{U_0(s)} = \frac{P(s)B}{b_0 B + \beta_3(s^2 + as)P(s) - b_0\beta_3}, \tag{10.6}$$

where $B = s^3 + (a+\beta_1)s^2 + (a\beta_1 + \beta_2)s + \beta_3$.

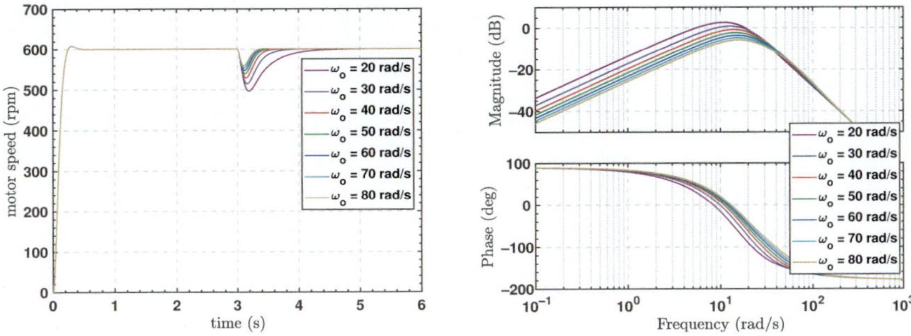

(a) Speed responses with different ω_o. **(b)** Sensitivity Bode plots with different ω_o.

FIGURE 10.2 Speed responses and sensitivity Bode plots with different ω_o.

Substituting $P(s) = \frac{b}{s(s+a)}$ into (10.6) yields

$$P_s(s) = \frac{Y(s)}{U_0(s)} = \frac{1}{s(s+a)}. \tag{10.7}$$

It can be seen from Eq. (10.7) that the simplified plant has nothing related to observer gain L, which means that the tracking performance is only related to the FOPD controller. Meanwhile, ω_o only affects the anti-load disturbance performance. In order to verify this result, given a certain FOPD controller with different ω_o, the step responses simulation is performed. The transfer function of the PMSM speed servo plant is shown in Eq. (10.8) using the output-error-based modeling method. The simulation results are shown in Figure 10.2a.

$$P(s) = \frac{383.635}{s(s+26.08)}. \tag{10.8}$$

In Figure 10.2a, the speed-tracking performances are exactly the same, although the values of GESO bandwidth ω_o are different, which means the tracking performance is only related to the FOPD controller. Furthermore, the anti-load disturbance performance is positively correlated with ω_o with the FOPD controller. In summary, it is verified that the FOPD-GESO controller meets the separation principle. Therefore, the speed-tracking response can be optimized by satisfying proper frequency domain and time-domain specifications, and the anti-load disturbance robustness can be adjusted by GESO parameter tuning, respectively.

10.2.3 Disturbance Rejection Analysis

In this subsection, the sensitive transfer function of disturbance d to tracking error e is given, which can be used as disturbance rejection analysis.

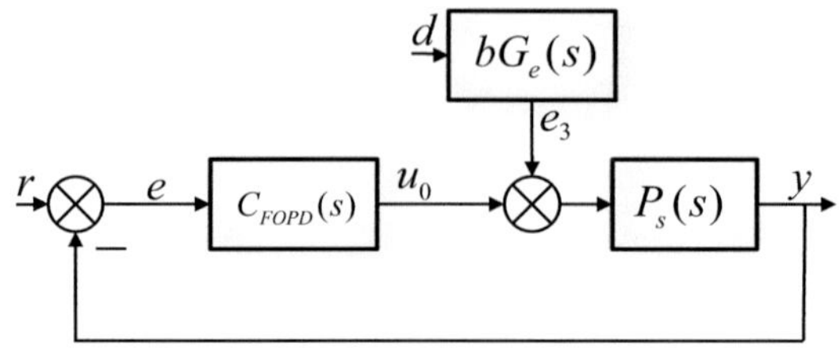

FIGURE 10.3 Block diagram of the FOPD-GESO control system.

By subtracting (10.1) from (9.1), the error dynamic is obtained as

$$\begin{bmatrix} \dot{e}_1 \\ \dot{e}_2 \\ \dot{e}_3 \end{bmatrix} = \begin{bmatrix} -\beta_1 & 1 & 0 \\ -\beta_2 & -a & 1 \\ -\beta_3 & 0 & 0 \end{bmatrix} \begin{bmatrix} e_1 \\ e_2 \\ e_3 \end{bmatrix} + \begin{bmatrix} 0 \\ 0 \\ 1 \end{bmatrix} h, \tag{10.9}$$

where $e_1 = x_1 - z_1$, $e_2 = x_2 - z_2$, and $e_3 = x_3 - z_3$.

Based on (10.9), the transfer function of disturbance estimation error is

$$G_e(s) = \frac{e_3(s)}{f(s)} = \frac{s^3 + (\beta_1 + a)s^2 + (\beta_2 + a\beta_1)s}{s^3 + (\beta_1 + a)s^2 + (\beta_2 + a\beta_1)s + \beta_3}. \tag{10.10}$$

According to the Figure 10.1, the input of the speed servo system can be presented as

$$u_1 = \frac{u_0 - z_3}{b_0} + d = \frac{u_0 + e_3}{b_0}. \tag{10.11}$$

Note $P(s) = bP_s(s)$ and $f = bd(t)$, combining Eqs. (10.10) and (10.11), then the control block diagram of FOPD-GESO feedback control system can be represented as shown in Figure 10.3.

Thus, the sensitive transfer function of disturbance d to tracking error e can be obtained as

$$E(s) = \frac{e(s)}{d(s)} = \frac{bG_e(s)P_s(s)}{1 + C_{FOPD}(s)P_s(s)}. \tag{10.12}$$

The sensitivity Bode plots corresponding to the examples in Figure 10.2a are shown in Figure 10.2b, which shows that the anti-load disturbance performance is correlated with ω_o.

10.3 PARAMETERS TUNING PROCEDURE

In this chapter, two frequency-domain specifications, i.e., gain crossover frequency and phase margin, and a time-domain specification, i.e., integrated time absolute error (ITAE), are applied to design the FOPD-ESO controller systematically. The considered feedback control system is shown in Figure 10.4. $P_s(s)$ is the simplified

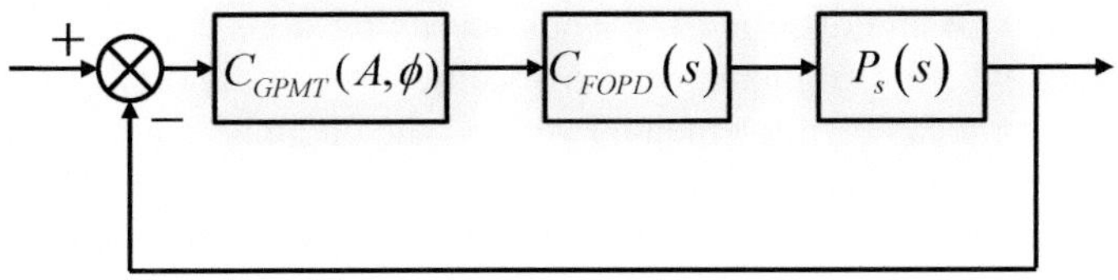

FIGURE 10.4 Feedback control system with the a simplified plant.

plant, and $C_{FOPD}(s)$ is the designed FOPD in the FOPD-GESO controller. There are four tuning parameters: K_p, K_d, μ, and ω_o.

In Figure 10.4, GMPT is a gain-phase margin tester.

The closed-loop transfer function of the control system in Figure 10.4 is

$$G(s) = \frac{M_T(A,\phi)C_{FOPD}(s)P_s(s)}{1+M_T(A,\phi)C_{FOPD}(s)P_s(s)}. \tag{10.13}$$

Accordingly, the characteristic equation of the closed-loop system (10.13) is

$$D(s;K_p,K_d,\mu,A,\phi) = s(s+a)+Ae^{-j\phi}(K_p+K_d s^{\mu}) = 0. \tag{10.14}$$

Substituting $j\omega$ for s in (10.14) yields

$$\begin{aligned} &D(s=j\omega;K_p,K_d,\mu,A,\phi),\\ &= j\omega(j\omega+a)+Ae^{-j\phi}(K_p+K_d(j\omega)^{\mu}) = 0. \end{aligned} \tag{10.15}$$

The real part and imaginary part of (10.15) can be obtained,

$$-\omega^2+AC_1E+AS_1F = 0, \tag{10.16}$$

$$a\omega+AC_1F-AS_1E = 0, \tag{10.17}$$

where

$$S_1 = sin\phi,\ C_1 = cos\phi, S_2 = sin\frac{\mu\pi}{2},\ C_2 = cos\frac{\mu\pi}{2},$$
$$E = K_p+K_d\omega^{\mu}C_2, F = K_d\omega^{\mu}S_2.$$

By solving equations, K_d and K_p concerning μ, ω, and ϕ can be expressed as

$$K_d = \frac{\omega^2 S_1 - aC_1\omega}{A\omega^{\mu}S_2}, \tag{10.18}$$

$$K_p = \frac{\omega^2-(AC_1K_d\omega^{\mu}C_2+AS_1K_d\omega^{\mu}S_2)}{AC_1}. \tag{10.19}$$

Given a fixed ϕ, μ, and ω, the other two parameters K_p and K_d can be determined a point with $A=1$ according Eqs. (10.18) and (10.19). This point can satisfy the pre-specified $\phi_m = \phi$ and $\omega_c = \omega$ [33]. Thus, sweeping $\mu \in (0,2)$, a series of points can be obtained which can guarantee two specifications ω_c and ϕ_m. Then, the step response simulation can be implemented with all the obtained points, and the corresponding J_{ITAE} can be calculated using Eq. (3.14) with a fixed ω_o. The parameters corresponding to the smallest J_{ITAE} can be determined. At last, the excellent anti-load disturbance ability can be achieved with proper ω_o.

According to the above analysis, the PMSM speed control plant can always be simplified to $\frac{1}{s(s+a)}$ through the GESO. Therefore, given gain crossover frequency and phase margin, the FOPD controller can be designed with a certain μ. Sweeping all the $\mu \in (0,2)$, all the FOPD controllers satisfying prespecified gain crossover frequency and phase margin can be obtained. Then, J_{ITAE} corresponding to the previously obtained parameters can be calculated through simulation. Select the parameters corresponding to the smallest J_{ITAE} as the final designed FOPD controller. Since the designed FOPD-GESO controller meets the separation principle, it is possible to give a suitable phase margin to achieve a small overshoot, and the anti-load disturbance performance can be adjusted by the parameter ω_o of GESO. The choice of ω_o is limited by the system noise, so it should be selected reasonably. The parameter setting rules of FOPD-GESO are summarized with an example as follows:

1. Considering the previously described PMSM speed servo plant, the frequency-domain specifications yield gain crossover frequency $\omega_c = 10\ \mathrm{rad/s}$ and phase margin $\phi_m = 70°$.

2. Given a fixed μ, the parameters of the FOPD controller can be determined according to Eqs. (10.18) and (10.19). Sweeping all the $\mu \in (0,2)$, all the FOPD controllers satisfying the prespecified ω_c and ϕ_m can be obtained and shown in Figure 10.5a.

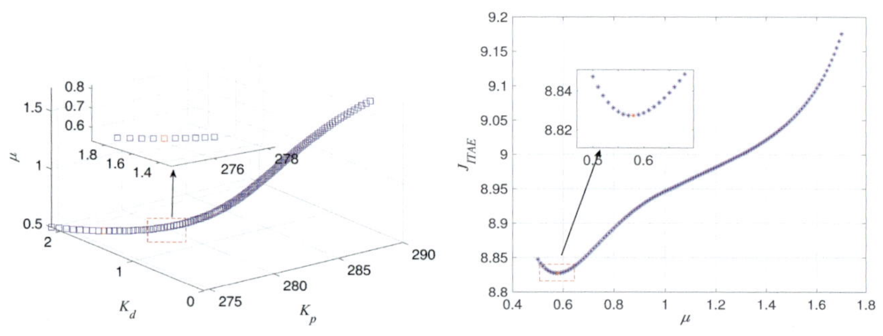

(a) All the parameters satisfying pre-specified ω_o and ϕ_m. **(b)** J_{ITAE}.

FIGURE 10.5 All the parameters satisfying prespecified ω_o and ϕ_m and corresponding J_{ITAE}.

3. Given a fixed ω_o, selecting every point obtained above, the step response simulations are implemented and the J_{ITAE} can be calculated. The correspondence diagram between μ and J_{ITAE} is presented as shown in Figure 10.5b. The smallest J_{ITAE} corresponding to the $\mu = 0.58$ is shown as a red star in Figure 10.5b. The parameters (K_p, K_d, μ) of the optimal FOPD-GESO controller can be obtained as Eq. (10.20).

$$C_{FOPD-GESO1}(s) = 275.5735 + 1.5880s^{0.58}. \tag{10.20}$$

4. Adding step disturbance before the PMSM speed servo plant, adjust ω_o to observe the anti-load disturbance response and finally choose $\omega_o = 300$ rad/s.

10.4 STABILITY ANALYSIS

The theoretical stability proof is addressed in this subsection: First, we give a proof of the stability of GESO; then, we give the stability boundary of the entire closed-loop controller to ensure the stability of the control system.

10.4.1 Stability of the GESO

According to (10.9), one can obtain

$$A_e = A - LC = \begin{bmatrix} -\beta_1 & 1 & 0 \\ -\beta_2 & -a & 1 \\ -\beta_3 & 0 & 0 \end{bmatrix}. \tag{10.21}$$

The characteristic polynomial of A_e is

$$\lambda(s) = s^3 + (a+\beta_1)s^2 + (a\beta_1 + \beta_2)s + \beta_3. \tag{10.22}$$

Place all the poles of the GESO to the same frequency of GESO bandwidth ω_o, i.e., $\lambda(s) = (s+\omega_o)^3$. Thus,

$$\beta_1 = 3\omega_o - a,\ \beta_2 = 3\omega_o^2 - a\beta_1, \beta_3 = \omega_o^3. \tag{10.23}$$

Obviously, the GESO is bounded-input bounded-output (BIBO) stable because all the roots of the characteristic polynomial are in the left plane and d is bounded.

10.4.2 Stability of the Closed-Loop System

Theorem 10.1

If choosing proper parameters K_p, K_d, and μ to ensure

$$|arg(w_i)| > \lambda\frac{\pi}{2}, \tag{10.24}$$

where w_i are the ith root of the following equation:

$$w^{2pq} + aw^{pq} + K_d w^{p^2} + K_p = 0, \tag{10.25}$$

where $\lambda = 1/pq$, $p/q = \mu$, p, and q are positive integers, then the closed-loop system is bounded, as $t \to \infty$. ■

Proof. The error E(s) can be obtained,

$$E(s) = R(s) - E(s)C_{FOPD}(s)P_s(s), \tag{10.26}$$

which is

$$E(s) = \frac{R(s)}{1 + C_{FOPD}(s)P_s(s)}. \tag{10.27}$$

The characteristic equation of (10.27) is

$$N(s) = s^2 + as + K_d s^\mu + K_p. \tag{10.28}$$

Notate that

$$\mu = \frac{p}{q},\ s^\mu = s^{p/q}, \tag{10.29}$$

where q and p are positive integers. So,

$$N(s) = s^2 + as + K_d s^{p/q} + K_p. \tag{10.30}$$

Notate $\lambda = 1/pq$, $w = s^\lambda$, then

$$N(w) = w^{2pq} + aw^{pq} + bK_d w^{p^2} + bK_p. \tag{10.31}$$

According to Lemma 6.1, $\overline{m} = \overline{n} = 1$, since,

$$|arg(w_i)| > \lambda \frac{\pi}{2}. \tag{10.32}$$

Therefore, the closed-loop system is BIBO. □

10.5 SIMULATION RESULTS

The PMSM speed can be controlled in a MATLAB simulation, using the proposed FOPD-GESO controller. The advantages are clearly shown through comparison with the typical IOPID, IOADRC, and FOPID controllers.

10.5.1 Comparison with the IOPID Controller

For fair comparison, the IOPID controller can be designed with the same $\omega_c = 10$ rad/s, $\phi_m = 70^\circ$ and the flat phase constraint [125],

$$C_{IOPID}(s) = 0.7280 + \frac{1.1757}{s} + 0.013s. \tag{10.33}$$

The open-loop, closed-loop, and sensitivity Bode plots of two control systems are shown in Figures 10.6, 10.7a, and 10.7b, which show that the two control systems

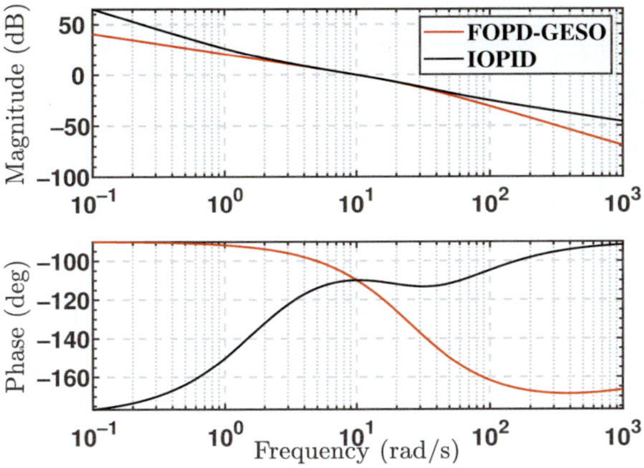

FIGURE 10.6 Open-loop Bode plots of two control systems.

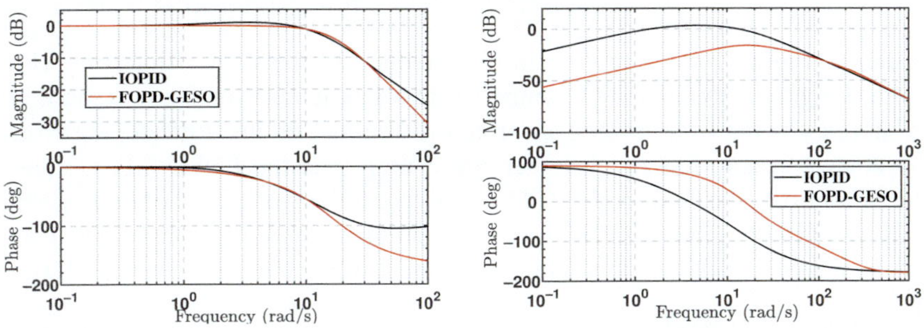

(a) Closed-loop Bode plots. **(b)** Sensitivity Bode plots.

FIGURE 10.7 Closed-loop and sensitivity Bode plots of two control systems.

both satisfy the given gain crossover frequency and phase margin. However, the designed FOPD-GESO control system has smaller resonance peaks than the IOPID control system, which means that the maximum overshoot of FOPD-GESO control system is smaller. Moreover, according to Figure 10.7b, the designed FOPD-GESO control system has better disturbance rejection performance than the IOPID control system.

10.5.1.1 Step Response and Anti-Load Disturbance Response

Setting the reference speed as 600 rpm, the speed step response simulations are performed, using the FOPD-GESO controller $C_{FOPD-GESO1}(s)$ and typical the IOPID controller $C_{IOPID}(s)$. The simulation results, step responses and control signals are shown in Figure 10.8.

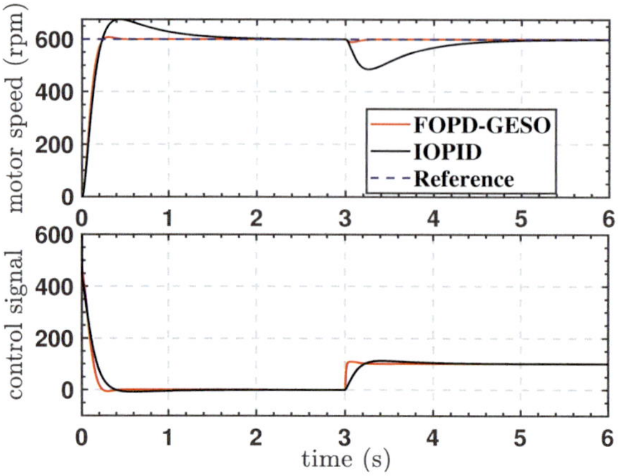

FIGURE 10.8 The speed responses (solid) and control signals (dotted) of FOPD-GESO and IOPID control systems (simulation).

According to Figure 10.8, the two control systems have the same rising time. However, the designed FOPD-GESO control system has much smaller overshoot and much shorter settling time. Moreover, in terms of anti-load disturbance, the typical IOPID control system has much larger speed drop than the FOPD-GESO control system. Therefore, the designed FOPD-GESO controller not only achieves desired tracking performance but also achieves outstanding anti-load disturbance, which breaks the inherent trade-off on tracking and anti-disturbance of the typical PID controller.

10.5.1.2 Robustness to Internal Uncertainties

In order to verify the robustness to internal uncertainties of the design FOPD-GESO controller, the gain of the PMSM speed servo plant is set to be two and four times the normal value for simulation. Step responses of the FOPD-GESO control system and typical IOPID control system are shown in Figure 10.9.

According to Figure 10.9, the overshoots of the responses of the FOPD-GESO control systems with gain variation have minor changes. For the typical IOPID control system, the overshoots and rising time of the responses with gain variation have obvious changes. Thus, the designed FOPD-GESO controller can achieve better robustness to internal uncertainties.

10.5.2 Comparison with the IOADR Controller

The proposed FOPD-GESO controller is also compared with the IOADRC. For fair comparison, the two controllers are designed with the same frequency-domain indicators and the same $\omega_o = 300$ rad/s.

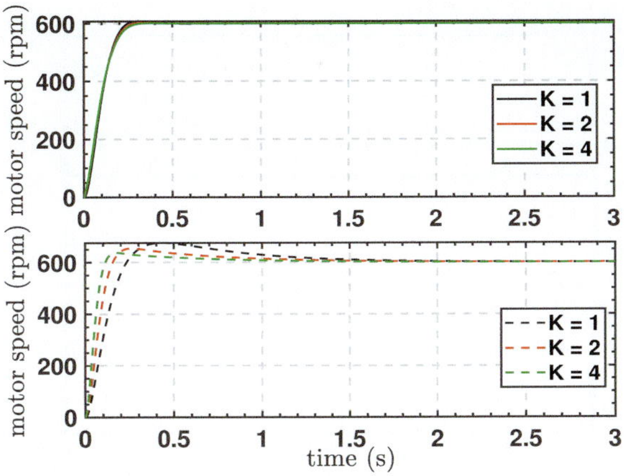

FIGURE 10.9 The speed responses of FOPD-GESO (solid) and IOPID (dotted) with different plant gains (simulation).

Given the frequency-domain specifications gain crossover frequency $\omega_c = 20$ rad/s and phase margin $\phi_m = 80°$, the FOPD-GESO controller can be designed as

$$C_{FOPD-GESO2}(s) = 564.045 + 17.132s^{0.96}. \tag{10.34}$$

Keeping the same frequency-domain indicators, the IOADRC can be designed as

$$C_{IOADRC}(s) = 100.855 + 24.638s. \tag{10.35}$$

Setting the reference speed as 600 rpm, the speed step response simulation is performed, using the designed $C_{FOPD-GESO2}(s)$ and $C_{IOADRC}(s)$. The simulation results of step responses and control signals are shown in Figure 10.10.

According to Figure 10.10, the designed FOPD-GESO control system has no overshoot and much shorter settling time. Furthermore, in terms of anti-load disturbance, the IOADRC control system has larger speed drop and much longer settling time than the FOPD-GESO control system.

10.5.3 Comparison with the FOPID Controller

The proposed FOPD-GESO controller is also compared with the FOPID controller [24]. Given the frequency specifications gain crossover frequency $\omega_{gc} =$ 15 rad/s and phase margin $\phi_m = 75°$, the FOPD-GESO controller can be designed with $\omega_o = 300$ rad/s,

$$C_{FOPD-GESO3}(s) = 421.44 + 9.687s^{0.92}. \tag{10.36}$$

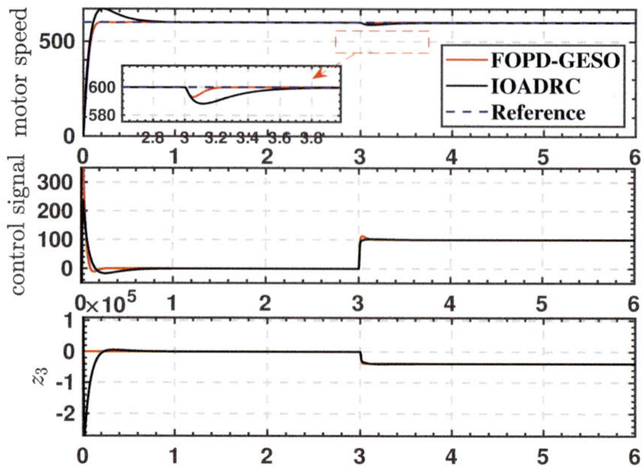

FIGURE 10.10 The speed responses, control signals, and z_3 of FOPD-GESO and IOADRC control systems (simulation).

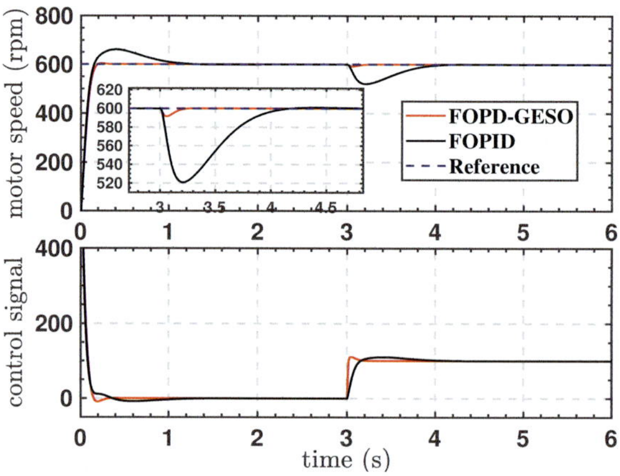

FIGURE 10.11 The speed responses (solid) and control signals (dotted) of FOPD-GESO and FOPID control systems (simulation).

Keeping the same frequency specifications, the FOPID controller can be designed as based on the proposed design method:

$$C_{FOPID}(s) = 0.78(1 + \frac{2.35}{s^{0.7245}} + 0.1063s^{0.7245}). \tag{10.37}$$

Setting the reference speed as 600 rpm, the speed step response simulations are performed with the designed FOPD-GESO and FOPID control systems. The step responses and control signals are shown in Figure 10.11.

According to Figure 10.11, two step responses have the same rising time. However, the designed FOPD-GESO control system response has much smaller overshoot and shorter settling time. Furthermore, the designed FOPID control system has much larger speed drop than the proposed FOPD-GESO control system for load disturbance response. Thus, the designed FOPD-GESO control system not only achieves better tracking response but also better anti-load disturbance rejection over the optimal FOPID control system.

10.6 EXPERIMENTAL VALIDATION

10.6.1 Comparison with the IOPID Controller

10.6.1.1 Step Response and Anti-Load Disturbance Response

To verify the speed-tracking performance and anti-load disturbance performance, the speed control experiment is performed. n_r is a step signal from 0 to 600 rpm. The speed responses are shown in Figure 10.12, using the designed $C_{FOPD-GESO}(s)$ controller and $C_{IOPID}(s)$ controller.

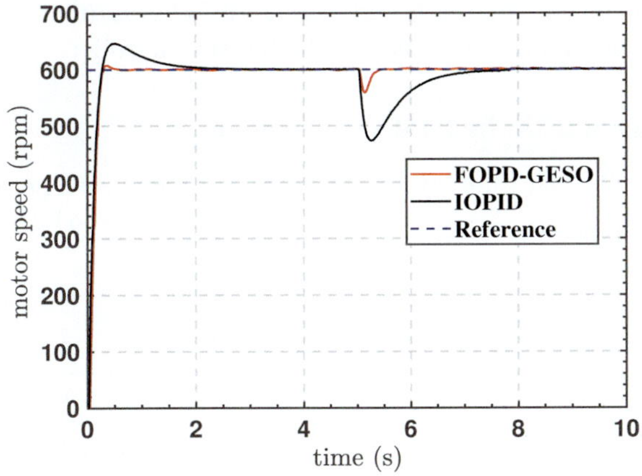

FIGURE 10.12 The speed responses of FOPD-GESO and IOPID control systems (experiment).

As shown in Figure 10.12, two control systems have the same rising time. However, the overshoot of the IOPID control system is 7.88%, and the setting time is 2.36 s. The FOPD-GESO controller shows the desired speed-tracking performance; the overshoot is 1.18% and the setting time is 0.50 s. Moreover, in terms of anti-load disturbance performance, the speed drop using the FOPD-GESO controller is 6.87%, and using the IOPID controller is 21.12%. The comparison results are shown in Table 10.1. Therefore, the designed FOPD-GESO controller achieves much better speed tracking and anti-load disturbance performance over IOPID.

TABLE 10.1
Comparison of Speed Responses with Two Control Systems (Experiment)

Controller	Overshoot (%)	Settling Time (s)	Speed Drop (%)
IOPID	7.88	2.36	21.12
FOPD-GESO	1.18	0.50	6.87

10.6.1.2 Robust to Internal Uncertainties

To show the robustness to the internal uncertainties, speed responses experiments with different gains are performed, using the two designed controllers. Setting the gain as one, two and four times the nominal value, the speed responses are shown in Figure 10.13.

According to Figure 10.13, the speed-tracking process is stable and the overshoots are close to each other for the FOPD-GESO control system with different gains; the overshoots and rising time are obviously different for the IOPID control system with different gains. Thus, the proposed FOPD-GESO controller achieves higher robustness to the internal uncertainties over the IOPID controller.

10.6.2 Comparison with the IOADR Controller

Setting the n_r as 600 rpm, the speed control experiments are performed, using the designed $C_{FOPD-GESO}(s)$ and $C_{IOPID}(s)$. The speed responses are shown in Figure 10.14.

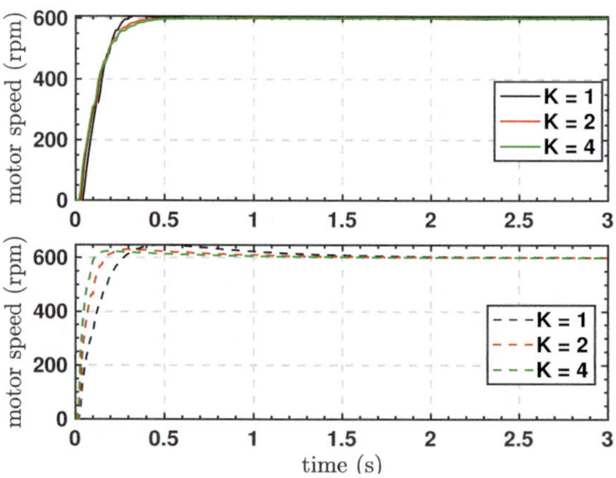

FIGURE 10.13 The speed responses of FOPD-GESO (solid) and IOPID (dotted) with different plant gains (experiment).

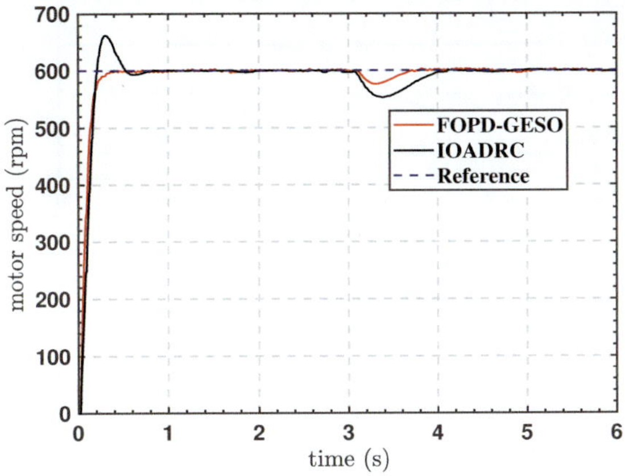

FIGURE 10.14 The speed responses of FOPD-GESO and IOADRC control systems (experiment).

TABLE 10.2
Comparison of Speed Responses with Two Control Systems (Experiment)

Controller	Overshoot (%)	Setting Time (s)	Speed Drop (%)
IOADRC	10.38	0.875	8.73
FOPD-GESO	0	0.41	4.0

As shown in Figure 10.14, the FOPD-GESO control system has no overshoot and the settling time is 0.41 s. The overshoot with IOADRC is 10.38% and the settling time is 0.875 s. Moreover, in terms of anti-load disturbance performance, the speed drop using the FOPD-GESO controller is 4.0%, and using IOADRC is 8.73%. The comparison results are shown in Table 10.2. Therefore, the designed FOPD-GESO controller achieves much better speed-tracking and anti-load disturbance performances than the IOADRC.

10.6.3 Comparison with the FOPID Controller

10.6.3.1 Tracking Step Signals

Setting the n_r as 600 rpm, the speed control experiments are performed, using the designed FOPD-GESO and FOPID controllers. The speed responses are shown in Figure 10.15.

As shown in Figure 10.15, the FOPD-GESO control system has no overshoot and the settling time is 0.29 s. The overshoot with the FOPID controller is 9.5% and the settling time is 0.92 s. Moreover, in terms of anti-load disturbance

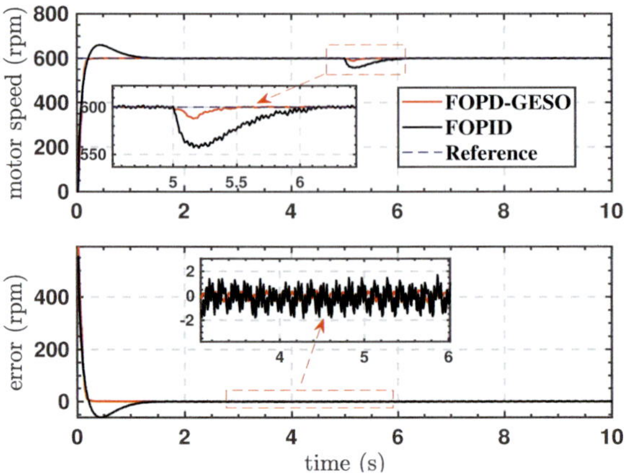

FIGURE 10.15 The speed responses of FOPD-GESO and FOPID control systems (experiment).

TABLE 10.3

Comparison of Speed Responses with Two Control Systems (Experiment)

Controller	Overshoot (%)	Setting Time (s)	Speed Drop (%)
FOPID	9.5	0.92	7.2
FOPD-GESO	0	0.29	2.05

performance, the speed drops of FOPD-GESO and FOPID are 2.05% and 7.2%, respectively. The comparison results are shown in Table 10.3. Thus, the designed FOPD-GESO controller achieves much better speed-tracking and anti-load disturbance performances than FOPID.

10.6.4 Extended Discussion

The memory allocation map of the microprocessor in Code Composer Studio (CCS) software showed that the proposed FOPD-GESO control algorithms occupies about 1.3 kbytes of the memory, which is far lower than the rated memory of the DSPTMS320F28335 (256k $\times$ 16 flash memory). Moreover, the comparative analysis in terms of computational complexity is presented. The execution time and utilized computational resources of Digital Signal Processing (DSP) for the four control methods are shown in Table 10.4. It is observed that the execution time of the proposed method is about 11.193 microseconds. Because the sampling frequency of the speed loop is 1.6 kHz, only 1.791% of the computational resources of DSP will be utilized. Compared to other control methods, the proposed method requires greater resources. However, the execution time is completely acceptable for the sampling time of the speed loop.

TABLE 10.4
Comparison of Computational Complexity

Control Method	FOPD-GESO	IOPID	IOADRC	FOPID
Execution time (μs)	11.193	1.707	9.907	6.980

10.7 CONCLUSION

In this chapter, a FOPD-GESO controller including a FOPD and a GESO is proposed for a typical PMSM speed motion system. The designed FOPD-GESO controller meets the separation principle. Thus, the FOPD controller achieves desired set-point tracking performance, and GESO achieves outstanding anti-load disturbance performance, which breaks the inherent trade-off on set-point tracking and anti-load disturbance of the typical PID control. The systematic design scheme of the FOPD-GESO controller for the PMSM speed servo system is presented with an illustration of an example, satisfying both frequency-domain and time-domain specifications. Simulation and experiments are carried out. Compared with a typical PID, IOADRC, and FOPID controllers, the designed FOPD-GESO controller can achieve much better speed tracking and the load disturbance rejection performance. Furthermore, it can achieve much better robustness to internal uncertainties than the typical PID controller.

References

1. Concepción A Monje, YangQuan Chen, Blas M Vinagre, Dingyu Xue, and Vicente Feliu-Batlle. *Fractional-order systems and controls: fundamentals and applications*. Springer Science & Business Media, 2010.
2. Denis Matignon. Generalized fractional differential and difference equations: Stability properties and modelling issues. *In Proceedings of Math. Theory of Networks and Systems Symposium*, 1998.
3. Hendrik W. Bode. *Network analysis and feedback amplifier design*. D. Van Nostrand Co. Inc, 1945.
4. S. Manabe. The non-integer integral and its application to control systems. *The Journal of the Institute of Electrical Engineers of Japan*, 80(860):589–597, 1960.
5. Boris J. Lurie. Three-parameter tunable tilt-integral-derivative (tid) controller. *US Patent US5371670*, 1994.
6. A. Oustaloup and M. Nouillant. Nouveau systéme de suspension: La suspension crone. *INPI Patent 90 046*, 13, 1990.
7. Alain Oustaloup, Benoit Mathieu, and Patrick Lanusse. Second generation crone control. In *Proceedings of IEEE Systems Man and Cybernetics Conference-SMC*, pp. 136–142. IEEE, 1993.
8. Patrick Lanusse, Alain Oustaloup, and Benoit Mathieu. Third generation crone control. In *Proceedings of IEEE Systems Man and Cybernetics Conference-SMC*, pp. 149–155. IEEE, 1993.
9. Igor Podlubny. Fractional-order systems and $\mathrm{PI}^{\lambda}\mathrm{D}^{\mu}$ controllers. *IEEE Transactions on Automatic Control*, 44(1):208–214, 1999.
10. Ivo Petras. The fractional-order controllers: Methods for their synthesis and application. *arXiv preprint math/0004064*, 2000.
11. Y. Luo, T. Zhang, B. Lee, C. Kang, and Y. Chen. Fractional-order proportional derivative controller synthesis and implementation for hard-disk-drive servo system. *IEEE Transactions on Control Systems Technology*, 22(1):281–289, 2014.
12. Ying Luo and YangQuan Chen. Fractional order [proportional derivative] controller for a class of fractional order systems. *Automatica*, 45(10):2446–2450, 2009.
13. Jingqing Han. From PID to active disturbance rejection control. *IEEE Transactions on Industrial Electronics*, 56(3):900–906, 2009.
14. Zhiqiang Gao. Scaling and bandwidth-parameterization based controller tuning. In *Proceedings of the American Control Conference*, pp. 4989–4996. IEEE, 2006.

15. HongSheng Li, Ying Luo, and YangQuan Chen. A fractional order proportional and derivative (FOPD) motion controller: tuning rule and experiments. *IEEE Transactions on Control Systems Technology*, 18(2):516–520, 2009.

16. Ying Luo, Yang Quan Chen, Chun Yang Wang, and You Guo Pi. Tuning fractional order proportional integral controllers for fractional order systems. *Journal of Process Control*, 20(7):823–831, 2010.

17. ChunYang Wang, Ying Luo, and YangQuan Chen. Fractional order proportional integral (FOPI) and [proportional integral] (FO[PI]) controller designs for first order plus time delay (FOPTD) systems. In *Proceedings of the 2009 Chinese Control and Decision Conference*, pp. 329–334. IEEE, 2009.

18. Weijia Zheng and Youguo Pi. Study of the fractional order proportional integral controller for the permanent magnet synchronous motor based on the differential evolution algorithm. *ISA Transactions*, 63:387–393, 2016.

19. Arijit Biswas, Swagatam Das, Ajith Abraham, and Sambarta Dasgupta. Design of fractional-order $PI^{\lambda}D^{\mu}$ controllers with an improved differential evolution. *Engineering Applications of Artificial Intelligence*, 22(2):343–350, 2009.

20. Indranil Pan and Saptarshi Das. Fractional order fuzzy control of hybrid power system with renewable generation using chaotic PSO. *ISA Transactions*, 62:19–29, 2016.

21. Alamdar Ravari Mahdieh and Yaghoobi Mahdi. Optimum design of fractional order PID controller using chaotic firefly algorithms for a control CSTR system. *Asian Journal of Control*, 21:2245–2255, 2019.

22. Ying Luo, YangQuan Chen, and Youguo Pi. Experimental study of fractional order proportional derivative controller synthesis for fractional order systems. *Mechatronics*, 21(1):204–214, 2011.

23. Blas M. Vinagre, Igor Podlubny, L. Dorcak, and V. Feliu. On fractional PID controllers: a frequency domain approach. *IFAC Proceedings Volumes*, 33(4):51–56, 2000.

24. Amélie Chevalier, Clovis Francis, Cosmin Copot, Clara M. Ionescu, and Robin De Keyser. Fractional-order pid design: Towards transition from state-of-art to state-of-use. *ISA Transactions*, 84:178–186, 2019.

25. Weijia Zheng, Ying Luo, YangQuan Chen, and Xiaohong Wang. A simplified fractional order PID controller's optimal tuning: A case study on a PMSM speed servo. *Entropy*, 23(2):130, 2021.

26. Wei Tang, Qing-Guo Wang, Xiang Lu, and Zhiping Zhang. Why Ti=4Td for PID controller tuning. In *Proceedings of the 2006 9th International Conference on Control, Automation, Robotics and Vision*, pp. 1–2. IEEE, 2006.

27. Jose Luis Guzman, Karl Johan Astrom, Sebastian Dormido, Tore Hagglund, Manuel Berenguel, and Yves Piguet. Interactive learning modules for PID control. *IEEE Control Systems Magazine*, 28(5):118–134, 2008.

28. Chang-Shing Tao, Che-Hsu Chang, and Kuang-Wei Han. Gain margins and phase margins for multivariable control systems with adjustable parameters. *International Journal of Control*, 54(2):435–452, 1991.

29. Serdar Ethem Hamamci. An algorithm for stabilization of fractional-order time delay systems using fractional-order PID controllers. *IEEE Transactions on Automatic Control*, 52(10):1964–1969, 2007.

30. Jürgen Ackermann and Dieter Kaesbauer. Stable polyhedra in parameter space. *Automatica*, 39(5):937–943, 2003.

31. Norbert Hohenbichler and Jürgen Ackermann. Synthesis of robust PID controllers for time delay systems. In *Proceedings of the 2003 European Control Conference (ECC)*, pp. 1169–1174. IEEE, 2003.

32. Chyi Hwang and Yi-Cheng Cheng. A numerical algorithm for stability testing of fractional delay systems. *Automatica*, 42(5):825–831, 2006.

33. Ying Luo and YangQuan Chen. Stabilizing and robust fractional order PI controller synthesis for first order plus time delay systems. *Automatica*, 48(9):2159–2167, 2012.

34. Yang Quan Chen and Kevin L. Moore. Relay feedback tuning of robust PID controllers with iso-damping property. *IEEE Transcations on Cybernetics*, 35(1):23–31, 2005.

35. K. V. Ling and A. L. Dexter. Expert control of air-conditioning plant. *Automatica*, 30(5):761–773, 1994.

36. Keqin Gu and S.-I. Niculescu. Survey on recent results in the stability and control of time-delay systems. *Journal of Dynamic Systems Measurement & Control*, 125:158–165, 2003.

37. Zhenlong Wu, Jie Yuan, Donghai Li, Yali Xue, and YangQuan Chen. The influence of rate limit on proportional–integral controller for first-order plus time-delay systems. *ISA Transactions*, 105:157–173, 2020.

38. Omar Waleed Abdulwahhab. Design of a complex fractional order PID controller for a first order plus time delay system. *ISA Transactions*, 99:154–158, 2020.

39. Aidan O'Dwyer. *Handbook of PI and PID controller tuning rules*. World Scientific, 2006.

40. Devendra Kumar and Dumitru Baleanu. Fractional calculus and its applications in physics. *Frontiers in Physics*, 7:81, 2019.

41. C. Ionescu, A. Lopes, Dana Copot, J.A. Tenreiro Machado, and J.H.T. Bates. The role of fractional calculus in modeling biological phenomena: A review. *Communications in Nonlinear Science and Numerical Simulation*, 51:141–159, 2017.

42. YangQuan Chen, Ivo Petras, and Dingyu Xue. Fractional order control-A tutorial. In *Proceedings of the 2009 American Control Conference*, pp. 1397–1411. IEEE, 2009.

43. Concepcion A. Monje, Antonio J. Calderon, Blas M. Vinagre, YangQuan Chen, and Vicente Feliu. On fractional PI controllers: Some tuning rules for robustness to plant uncertainties. *Nonlinear Dynamics*, 38(1-4):369–381, 2004.

44. Alain Oustaloup and Pierre Melchior. The great principles of the CRONE control. In *Proceedings of IEEE Systems Man and Cybernetics Conference-SMC*, pp. 118–129. IEEE, 1993.

45. Ying Luo and Yang Chen. *Fractional order motion controls*. John Wiley & Sons, 2012.

46. Tahmine V. Moghaddam, N. Bigdeli, and K. Afshar. Tuning a fractional order PID controller with lead compensator in frequency domain. *World Academy of Science, Engineering and Technology*, 75:1278–1283, 2011.

47. Erhan Yumuk, Müjde Güzelkaya, and İbrahim Eksin. Analytical fractional PID controller design based on Bode's ideal transfer function plus time delay. *ISA Transactions*, 91:196–206, 2019.

48. Helem Sabina Sánchez, Fabrizio Padula, Antonio Visioli, and Ramon Vilanova. Tuning rules for robust FOPID controllers based on multi-objective optimization with FOPDT models. *ISA Transactions*, 66:344–361, 2017.

49. Rahmat Khezri, Arman Oshnoei, Mehrdad Tarafdar Hagh, and SM Muyeen. Coordination of heat pumps, electric vehicles and AGC for efficient LFC in a smart hybrid power system via SCA-based optimized FOPID controllers. *Energies*, 11(2):420, 2018.

50. Tulasichandra Sekhar Gorripotu, Halini Samalla, Jagan Mohana Rao, Ahmad Taher Azar, Danilo Pelusi, et al. TLBO algorithm optimized fractional-order PID controller for AGC of interconnected power system. In *Soft Computing in Data Analytics*, pages 847–855. Springer, 2019.

51. Duarte Valério and José Sá Da Costa. Tuning of fractional PID controllers with Ziegler–Nichols-type rules. *Signal Processing*, 86(10):2771–2784, 2006.

52. Sigurd Skogestad. Simple analytic rules for model reduction and PID controller tuning. *Journal of Process Control*, 13(4):291–309, 2004.

53. Ju I. Neimark. D-decomposition of the space of quasi-polynomials (on the stability of linearized distributive systems). *American Mathematical Society Translations*, 102:95–131, 1973.

54. Pengchong Chen, Ying Luo, WeiJia Zheng, Zhiqiang Gao, and YangQuan Chen. Fractional order active disturbance rejection control with the idea of cascaded fractional order integrator equivalence. *ISA Transactions*, 114:359–369, 2021.

55. Sebastian Dormido. Advanced PID control. *IEEE Control Systems Magazine*, 26(1):98–101, 2006.

56. Karl Johan Áström and Panganamala Ramana Kumar. Control: A perspective. *Automatica*, 50(1):3–43, 2014.

57. Karl J Åström and Tore Hägglund. *PID controllers: theory, design, and tuning*, volume 2. Instrument Society of America Research Triangle Park, 1995.

58. Khuen Ho Weng, Chieh Hang Chang, and Lisheng S. Cao. Tuning of PID controllers based on gain and phase margin specifications. *Automatica*, 31(3):497–502, 1995.

59. Z. Shafiei and A. T. Shenton. Frequency-domain design of PID controllers for stable and unstable systems with time delay. *Automatica*, 33(12):2223–2232, 1997.

60. Ho Wang Fung, Qing Guo Wang, and Tong Heng Lee. PI tuning in terms of gain and phase margins. *Automatica*, 34(9):1145–1149, 1998.

61. Qing-Guo Wang, Ho-Wang Fung, and Yu Zhang. PID tuning with exact gain and phase margins. *ISA Transactions*, 38(3):243–249, 1999.

62. Li Sun, Donghai Li, and Kwang Y Lee. Optimal disturbance rejection for PI controller with constraints on relative delay margin. *ISA Transactions*, 63:103–111, 2016.

63. Zhenlong Wu, Donghai Li, and Yali Xue. A new pid controller design with constraints on relative delay margin for first-order plus dead-time systems. *Processes*, 7(10):713, 2019.

64. Sergei S. Mikhalevich, Sergey A. Baydali, and Flavio Manenti. Development of a tunable method for PID controllers to achieve the desired phase margin. *Journal of Process Control*, 25:28–34, 2015.

65. Hamideh Najafizadegan, Farshad Merrikh-Bayat, and Abolfazl Jalilvand. IMC-PID controller design based on loop shaping via LMI approach. *Chemical Engineering Research & Design*, 124:170–180, 2017.

66. Radek Matusu, Bilal Senol, and Libor Pekar. Robust PI control of interval plants with gain and phase margin specifications: Application to a continuous stirred tank reactor. *IEEE Access*, 8:145372–145380, 2020.

67. Chunna Zhao, Dingyu Xue, and YangQuan Chen. A fractional order PID tuning algorithm for a class of fractional order plants. In *IEEE International Conference Mechatronics and Automation, 2005*, pp. 216–221. IEEE, 2005.

68. Concepción A. Monje, Blas M. Vinagre, Vicente Feliu, and YangQuan Chen. Tuning and auto-tuning of fractional order controllers for industry applications. *Control Engineering Practice*, 16(7):798–812, 2008.

69. Ying Luo and Yangquan Chen. Synthesis of robust PID controllers design with complete information on pre-specifications for the FOPTD systems. In *Proceedings of the 2011 American Control Conference*, pp. 5013–5018, 2011.

70. Yangquan Chen. Impulse response invariant discretization of fractional order integrators/differentiators, filter design and analysis, MATLAB central.

`http://www.mathworks.com/matlabcentral/fileexchange/21342-impulse-response-invariant-discretization-of-fractional-order-/integrators-differentiator`, 2021.

71. Fabrizio Padula and Antonio Visioli. Tuning rules for optimal PID and fractional-order PID controllers. *Journal of Process Control*, 21(1):69–81, 2011.

72. Hendrik W. Bode. *Network analysis and feedback amplifier design*. 1946.

73. QingGuo Wang, C. C. Hang, and XuePing Yang. Single-loop controller design via IMC principles. *Automatica*, 91(12):2041–2048, 2001.

74. Majid Zamani, Masoud Karimi-Ghartemani, Nasser Sadati, and Mostafa Parniani. Design of a fractional order PID controller for an AVR using particle swarm optimization. *Control Engineering Practice*, 17(12):1380–1387, 2009.

75. Hsue-Shen Tsien. *Engineering cybernetics*. McGraw-Hill, 1954.

76. Isaac M Horowitz. *Synthesis of feedback systems*. Elsevier, 2013.

77. Duncan McFarlane and Keith Glover. A loop-shaping design procedure using H-infinity-synthesis. *IEEE Transactions on Automatic Control*, 37(6):759–769, 1992.

78. Xiuxing Yin and Li Pan. Direct adaptive robust tracking control for 6 DOF industrial robot with enhanced accuracy. *ISA Transactions*, 72:178–184, 2018.

79. Araki Mituhiko and Taguchi Hidefumi. Two-Degree-of-Freedom PID Controllers. *International Journal of Control Automation & Systems*, 1:401–411, 2003.

80. Wen-Hua Chen, Jun Yang, Lei Guo, and Shihua Li. Disturbance-Observer-Based Control and Related Methods-An Overview. *IEEE Transactions on Industrial Electronics*, 63(2):1083–1095, 2016.

81. Kemin Zhou, John Comstock Doyle, Keith Glover, et al. *Robust and optimal control*, volume 40. Prentice Hall New Jersey, 1996.

82. Zhiqiang Gao. Active disturbance rejection control: a paradigm shift in feedback control system design. In *2006 American Control Conference*, pp. 7–pp. IEEE, 2006.

83. Y. X. Su, C. H. Zheng, and B. Y. Duan. Automatic disturbances rejection controller for precise motion control of permanent-magnet synchronous motors. *IEEE Transactions on Industrial Electronics*, 52(3):814–823, 2005.

84. Yuanqing Xia, Liu Bo, and Mengyin Fu. Active disturbance rejection control for power plant with a single loop. *Asian Journal of Control*, 14(1):239–250, 2012.

85. Wenchao Xue and Yi Huang. On performance analysis of ADRC for a class of MIMO lower-triangular nonlinear uncertain systems. *ISA Transactions*, 53(4):955–962, 2014.

86. Gao Zhe. Active disturbance rejection control for nonlinear fractional-order systems. *International Journal of Robust & Nonlinear Control*, 26(4):876–892, 2016.

87. Mingda Li, Donghai Li, Jing Wang, and Chunzhe Zhao. Active disturbance rejection control for fractional-order system. *ISA Transactions*, 52(3):365–374, 2013.

88. Dazi Li, Pan Ding, and Zhiqiang Gao. Fractional active disturbance rejection control. *ISA Transactions*, 62:109–119, 2016.

89. Xinxin Shi, YangQuan Chen, and Jiacai Huang. Application of fractional-order active disturbance rejection controller on linear motion system. *Control Engineering Practice*, 81:207–214, 2018.

90. Mukesh D. Patil, Vishwesh A. Vyawahare, and Manisha K. Bhole. A new and simple method to construct root locus of general fractional-order systems. *ISA Transactions*, 53(2):380–390, 2014.

91. Denis Matignon. Stability results for fractional differential equations with applications to control processing. *Computational Engineering in Systems Applications*, 2:963–968, 1996.

92. Chen Dai, Tianliang Guo, Jun Yang, and Shihua Li. A disturbance observer-based current-constrained controller for speed regulation of PMSM systems subject to unmatched disturbances. *IEEE Transactions on Industrial Electronics*, 68(1):767–775, 2020.

93. Qiang Tan, Xuzhen Huang, Liyi Li, and Mingyi Wang. Magnetic field analysis and flux barrier design for modular permanent magnet linear synchronous motor. *IEEE Transactions on Industrial Electronics*, 67(5):3891–3900, 2020.

94. Changliang Xia, Bingnan Ji, and Yan Yan. Smooth speed control for low-speed high-torque permanent-magnet synchronous motor using proportional–integral–resonant controller. *IEEE Transactions on Industrial Electronics*, 62(4):2123–2134, 2014.

95. Feifei Bu, Zhida Yang, Yu Gao, Zihao Pan, Tianyu Pu, Michele Degano, and Chris Gerada. Speed ripple reduction of direct-drive PMSM servo system at low-speed operation using virtual cogging torque control method. *IEEE Transactions on Industrial Electronics*, 68(1):160–174, 2021.

96. Qiaofen Zhang, Haohao Guo, Chen Guo, Yancheng Liu, and Dunzhi Chen. An adaptive proportional-integral-resonant controller for speed ripple suppression of PMSM drive due to current measurement error. *International Journal of Electrical Power & Energy Systems*, 129:106866, 2021.

97. Myoungho Kim, Seung-Ki Sul, and Junggi Lee. Compensation of current measurement error for current-controlled PMSM drives. *IEEE Transactions on Industry Applications*, 50(5):3365–3373, 2014.

98. Michael C. Harke, Juan Manuel Guerrero, Michael W. Degner, Fernando Briz, and Robert D. Lorenz. Current measurement gain tuning using high-frequency signal injection. *IEEE Transactions on Industry Applications*, 44(5):1578–1586, 2008.

99. H.-S. Jung, S.-H. Hwang, J.-M. Kim, C.-U. Kim, and C. Choi. Diminution of current-measurement error for vector-controlled AC motor drives. *IEEE Transactions on Industry Applications*, 42(5):1249–1256, 2006.

100. Jiadong Lu, Yihua Hu, Guipeng Chen, Zheng Wang, and Jinglin Liu. Mutual calibration of multiple current sensors with accuracy uncertainties in IPMSM drives for electric vehicles. *IEEE Transactions on Industrial Electronics*, 67(1):69–79, 2020.

101. Kwang-Woon Lee and Sang-Il Kim. Dynamic performance improvement of a current offset error compensator in current vector-controlled spmsm drives. *IEEE Transactions on Industrial Electronics*, 66(9):6727–6736, 2019.

102. Quoc Nam Trinh, Peng Wang, Yi Tang, Leong Hai Koh, and Fook Hoong Choo. Compensation of DC offset and scaling errors in voltage and current measurements of three-phase AC/DC converters. *IEEE Transactions on Power Electronics*, 33(6):5401–5414, 2018.

103. Weizhe Qian, S. K. Panda, and Jian-Xin Xu. Torque ripple minimization in PM synchronous motors using iterative learning control. *IEEE Transactions on Power Electronics*, 19(2):272–279, 2004.

104. Yunda Yan, Jun Yang, Zhenxing Sun, Chuanlin Zhang, Shihua Li, and Haoyong Yu. Robust speed regulation for PMSM servo system with multiple sources of disturbances via an augmented disturbance observer. *IEEE/ASME Transactions on Mechatronics*, 23(2):769–780, 2018.

105. P. Mattavelli and F.P. Marafao. Repetitive-based control for selective harmonic compensation in active power filters. *IEEE Transactions on Industrial Electronics*, 51(5):1018–1024, 2004.

106. Shan Chai, Liuping Wang, and Eric Rogers. A cascade MPC control structure for a PMSM with speed ripple minimization. *IEEE Transactions on Industrial Electronics*, 60(8):2978–2987, 2012.

107. Quoc Nam Trinh, Peng Wang, Yi Tang, and Fook Hoong Choo. Mitigation of DC and harmonic currents generated by voltage measurement errors and grid voltage distortions in transformerless grid-connected inverters. *IEEE Transactions on Energy Conversion*, 33(2):801–813, 2018.

108. Pengchong Chen and Ying Luo. A two-degree-of-freedom controller design satisfying separation principle with fractional-order PD and generalized ESO. *IEEE/ASME Transactions on Mechatronics*, 27(1):137–148, 2021.

109. Deepak Pullaguram, Sukumar Mishra, Nilanjan Senroy, and Monish Mukherjee. Design and tuning of robust fractional order controller for autonomous microgrid VSC system. *IEEE Transactions on Industry Applications*, 54(1):91–101, 2018.

110. Dongdong Zhao, Fei Li, Rui Ma, Guosheng Zhao, and Yigeng Huangfu. An unknown input nonlinear observer based fractional order PID control of fuel cell air supply system. *IEEE Transactions on Industry Applications*, 56(5):5523–5532, 2020.

111. Reetam Mondal and Jayati Dey. Performance analysis and implementation of fractional order 2-DOF control on cart-inverted pendulum system. *IEEE Transactions on Industry Applications*, 56(6):7055–7066, 2020.

112. Geethi Krishnan, Shantanu Das, and Vivek Agarwal. An online identification algorithm to determine the parameters of the fractional-order model of a supercapacitor. *IEEE Transactions on Industry Applications*, 56(1):763–770, 2020.

113. Pushkar Prakash Arya and Sohom Chakrabarty. A robust internal model-based fractional order controller for fractional order plus time delay processes. *IEEE Control Systems Letters*, 4(4):862–867, 2020.

114. Sarthak Chatterjee, Orlando Romero, and Sérgio Pequito. A separation principle for discrete-time fractional-order dynamical systems and its implications to closed-loop neurotechnology. *IEEE Control Systems Letters*, 3(3):691–696, 2019.

115. Hadi Malek, Sara Dadras, Chun Yin, and YangQuan Chen. Fractional order proportional-resonant controller. In *2018 Annual American Control Conference (ACC)*, pp. 3086–3091. IEEE, 2018.

116. Zhanqing Zhou, Changliang Xia, Yan Yan, Zhiqiang Wang, and Tingna Shi. Disturbances attenuation of permanent magnet synchronous motor drives using cascaded predictive-integral-resonant controllers. *IEEE Transactions on Power Electronics*, 33(2):1514–1527, 2018.

117. WeiJia Zheng, Ying Luo, YangQuan Chen, and YouGuo Pi. Fractional-order modeling of permanent magnet synchronous motor speed servo system. *Journal of Vibration and Control*, 22(9):2255–2280, 2016.

118. Zhenbao Pan, Fei Dong, Jiwen Zhao, Lijun Wang, Hui Wang, and Yinyi Feng. Combined resonant controller and two-degree-of-freedom PID controller for PMSLM current harmonics suppression. *IEEE Transactions on Industrial Electronics*, 65(9):7558–7568, 2018.

119. Michael Ruderman and Torsten Bertram. Variable proportional-integral-resonant (PIR) control of actuators with harmonic disturbances. In *2013 IEEE International Conference on Mechatronics (ICM)*, pp. 847–852. IEEE, 2013.

120. Alejandro G. Yepes, Francisco D. Freijedo, Óscar Lopez, and Jesús Doval-Gandoy. High-performance digital resonant controllers implemented with two integrators. *IEEE Transactions on Power Electronics*, 26(2):563–576, 2011.

121. Chen Boshi. *Electric drag automatic control system: motion control system*. Machinery Industry Press, 2003.

122. Hebertt Sira-Ramírez, Jesús Linares-Flores, Carlos García-Rodríguez, and Marco Antonio Contreras-Ordaz. On the control of the permanent magnet synchronous motor: an active disturbance rejection control approach. *IEEE Transactions on Control Systems Technology*, 22(5):2056–2063, 2014.

123. Youwu Du, Weihua Cao, and Jinhua She. Analysis and design of active disturbance rejection control with an improved extended state observer for systems with measurement noise. *IEEE Transactions on Industrial Electronics*, 70(1):855–865, 2023.

124. Pengchong Chen, Ying Luo, Weijia Zheng, and Zhiqiang Gao. A new active disturbance rejection controller design based on fractional extended state observer. In 2019 *Chinese control conference (CCC)*, pages 4276–4281. IEEE, 2019.

125. YangQuan Chen, ChuanHua Hu, and Kevin L. Moore. Relay feedback tuning of robust PID controllers with iso-damping property. pp. 2180–2185. IEEE, 2003.

126. Saurabh Srivastava and V. S. Pandit. A 2-Dof LQR based PID controller for integrating processes considering robustness/performance tradeoff. *ISA Transactions*, 71:426–439, 2017.

127. Pawel Nowak, Jacek Czeczot, and Tomasz Klopot. Robust tuning of a first order reduced active disturbance rejection controller. *Control Engineering Practice*, 74:44–57, 2018.

128. Hebertt Sira-Ramirez and Marco Antonio Oliver-Salazar. On the robust control of buck-converter dc-motor combinations. *IEEE Transactions on Power Electronics*, 28(8):3912–3922, 2013.

129. Zhiqiang Gao, Yi Huang, and Jingqing Han. An alternative paradigm for control system design. In *Proceedings of the 40th IEEE Conference on Decision and Control (Cat. No. 01CH37228)*, pp. 4578–4585. IEEE, 2001.

130. Qing-Chang Zhong, Alon Kuperman, and R. K. Stobart. Design of UDE-based controllers from their two-degree-of-freedom nature. *International Journal of Robust and Nonlinear Control*, 21(17):1994–2008, 2011.

131. Qun Zhu, Zhonggang Yin, Yanqing Zhang, Jianbo Niu, Yan Li, and Yanru Zhong. Research on two-degree-of-freedom internal model control strategy for induction motor based on immune algorithm. *IEEE Transactions on Industrial Electronics*, 63(3):1981–1992, 2016.

132. Hoi-Wai Chow and Norbert C. Cheung. Disturbance and response time improvement of submicrometer precision linear motion system by using modified disturbance com-

pensator and internal model reference control. *IEEE Transactions on Industrial Electronics*, 60(1):139–150, 2013.

133. Jorge Rivera Dominguez, Christian Mora-Soto, Susana Ortega-Cisneros, Juan José Raygoza Panduro, and Alexander G. Loukianov. Copper and core loss minimization for induction motors using high-order sliding-mode control. *IEEE Transactions on Industrial Electronics*, 59(7):2877–2889, 2012.

134. Xiaoguang Zhang, Lizhi Sun, Ke Zhao, and Li Sun. Nonlinear speed control for PMSM system using sliding-mode control and disturbance compensation techniques. *IEEE Transactions on Power Electronics*, 28(3):1358–1365, 2013.

135. Qibing Jin, Wu Cai, and Xinghan Du. Disturbance observer-based control with partially coupled inverse model for stable TITO processes with time delays. *ISA Transactions*, 110:148–159, 2021.

136. Ping Zhou, Tian-You Chai, and Jin-Hui Zhao. Dob design for nonminimum-phase delay systems and its application in multivariable MPC control. *IEEE Transactions on Circuits and Systems II: Express Briefs*, 59(8):525–529, 2012.

137. Yunfeng Fan and U-Xuan Tan. Design of a feedforward-feedback controller for a piezoelectric-driven mechanism to achieve high-frequency nonperiodic motion tracking. *IEEE/ASME Transactions on Mechatronics*, 24(2):853–862, 2019.

138. Li Sun, Yuqiong Zhang, Donghai Li, and Kwang Y. Lee. Tuning of active disturbance rejection control with application to power plant furnace regulation. *Control Engineering Practice*, 92:104122, 2019.

139. Qing Zheng, Lili Dong, Dae Hui Lee, and Zhiqiang Gao. Active disturbance rejection control for MEMS gyroscopes. In *2008 American Control Conference*, pp. 4425–4430. IEEE, 2008.

140. Zhenlong Wu, Donghai Li, Yali Xue, Liming Sun, Ting He, and Song Zheng. Modified active disturbance rejection control for fluidized bed combustor. *ISA Transactions*, 102:135–153, 2020.

141. Caifen Fu and Wen Tan. Tuning of linear ADRC with known plant information. *ISA Transactions*, 65:384–393, 2016.

142. Rong Zhou and Wen Tan. Analysis and tuning of general linear active disturbance rejection controllers. *IEEE Transactions on Industrial Electronics*, 66(7):5497–5507, 2018.

143. Vahid Badri and Mohammad Saleh Tavazoei. Achievable performance region for a fractional-order proportional and derivative motion controller. *IEEE Transactions on Industrial Electronics*, 62(11):7171–7180, 2015.

Index

For Product Safety Concerns and Information please contact our EU representative GPSR@taylorandfrancis.com Taylor & Francis Verlag GmbH, Kaufingerstraße 24, 80331 München, Germany